SpringerBriefs in Energy
Energy Analysis

Series Editor: Charles A.S. Hall

For further volumes:
http://www.springer.com/series/8903

Lianyong Feng • Yan Hu
Charles A.S. Hall • Jianliang Wang

The Chinese Oil Industry

History and Future

 Springer

Lianyong Feng
Faculty of Economics and Trade
School of Business Administration
China University of Petroleum
Beijing 102249, China

Yan Hu
School of Business Administration
China University of Petroleum
Beijing 102249, China

Charles A.S. Hall
Faculty of Environmental
& Forest Biology and Graduate Program
in Environmental Science
College of Environmental
Science and Forestry
State University of New York
Syracuse, NY 13210, USA

Jianliang Wang
School of Business Administration
China University of Petroleum
Beijing 102249, China

ISSN 2191-5520   ISSN 2191-5539 (electronic)
ISBN 978-1-4419-9409-7 ISBN 978-1-4419-9410-3 (eBook)
DOI 10.1007/978-1-4419-9410-3
Springer New York Heidelberg Dordrecht London

Library of Congress Control Number: 2012952093

Printed on acid-free paper

Springer is part of Springer Science+Business Media (www.springer.com)

About the Authors

Professor Lianyong Feng was born in 1966. He is an associate professor at the School of Business Administration in the University of Petroleum, Beijing. Presently, he is secretary of ASPO-China.

In 1988, he graduated and was awarded his Management Engineering Bachelor's degree from the University of Petroleum (East China). He continued his studies there and received his Master's degree 3 years later. Since April of 1991, he started teaching at this university for the next 6 years. In March of 1996, Dr. Feng furthered his studies at the Moscow Petroleum University, Russia and received his Doctoral degree on 10 October 1997. From 1997 to 1999, Dr. Feng worked in Kazakhstan for the transportation and distribution of crude oil. From July 1999 to October 2000, he worked for the crude oil importation from Russia. From October 2000 to September 2003, he worked as deputy director in the Strategy Research Office of Development Research Center of China's National Petroleum Corporation (CNPC).

In September of 2003, Dr. Feng returned to the University of Petroleum, Beijing and started his teaching and research career again. He teaches courses on "International Petroleum Economics," "Engineering Economics," and "Energy Economics." Dr. Feng has published more than 30 articles and monographs. He has also undertaken and completed over ten research projects related to the subjects on international petroleum economics and cooperation and energy economics.

Yan Hu is a PhD candidate at the program of Petroleum Engineering and Management at the China University of Petroleum (Beijing), where she studied for 7 years since 2004. She has been studying peak oil and energy system and management, and has published more than ten scholarly articles in Chinese and one in a Chinese book "Post Oil Age." In 2011, Hu was supported by China Scholarship Council (CSC) to study in the State University of New York (SUNY) for more than 1 year. Recently, her interests have turned towards analyzing the energy return on investment (EROI) which was defined by Charles Hall. She seized the opportunities offered by Hall to attend SUNY, attending the classes including "Energy System," "System Ecology," and "Biophysical Economy," and also publishing one paper in

Sustainability on "Analysis of the Energy Return on Investment (EROI) of the Huge Daqing Oil Field in China."

Professor Charlie Hall is a systems ecologist with strong interests in energy flows in natural systems and human society. He received his PhD from Dr. Howard Odum at the University of North Carolina at Chapel Hill in 1970. His work has involved streams, estuaries, and tropical forests, but has focused increasingly on human-dominated ecosystems in the USA and Latin America. He is best known for developing the concept of EROI, or energy return on investment, as it relates to, e.g., migrating fish and obtaining oil and gas. Hall's latest focus has been on developing an alternative approach to economics called biophysical economics, an attempt to understand human economies from a biophysical rather than just social perspective. He recently coauthored "Energy and the Wealth of Nations: Understanding the Biophysical Economy" with economist Kent Klitgaard.

Jianliang Wang is a PhD student of Petroleum Engineering and Management in China University of Petroleum (Beijing). His main research area is evaluation of oil and gas resources, modeling peak fossil fuels, and analysis on the impact of the potential shortage of oil and gas supply on China's economy. He has published nine peer-reviewed articles in journals, three papers in magazines, four papers in newspapers, and several papers in conferences. In September 2012, he will be supported by China Scholarship Council (CSC) to study in the Global Energy System of Uppsala University in Sweden for 1 year. The topic of his doctoral thesis is about forecasting world's fossil fuels production and analyzing their impact on future global climate change.

Preface

This small book represents an attempt to make the main features of the Chinese oil industry available in English. It was written by two professionals from the China University of Petroleum in rough English. Coauthor Charles Hall did his best to modify the English so that the meaning was clear to English-speaking readers while preserving a little of the flavor of Chinese ways of communicating.

What China does or does not do about petroleum will have an enormous impact on the future of global petroleum. The Chinese Economy is now the second largest in the world, and its use of oil, increasingly imported, will have major impacts on the rest of the world.

Beijing, China	Yan Hu
Beijing, China	Lianyong Feng
NY, USA	Charles A. S. Hall
Beijing, China	Jianliang Wang

Acknowledgements

This book has been supported by National Natural Science Foundation of China (NO. 71073173). We thank Chao Qi for assistance with the data analysis and editing, Xu Tang for providing some forecasting results, Jim Gray, Carlos Pascualli, and Rigo Melgar for excellent editing of words and sentences, Alex Poisson for some suggestions.

Contents

Acronyms

URR	Ultimate Recoverable Reserves
CESY	China Energy Statistic Yearbook
CNOOC	China National Offshore Oil Corporation
CNPC	China National Petrochemical Corporation
CSY	China Statistical Yearbook
EIA	U.S. Energy Information Administration
MLR	Ministry of Land and Resources
NBSC	National Bureau of Statistics of China
NDRC	National Development and Reform Commission
OCPC	Oil consumption per capita
OPEC	The Organization of Petroleum Exporting Countries
PRC	People's Republic of China
Sinochem	Sinochem Group
Sinopec	China National Petroleum & Chemical Corporation
SPC	Syrian Petroleum Corporation
UAE	The United Arab Emirates
Yanchang Petroleum Group	Shaanxi Yanchang Petroleum (Group) Corp. Ltd

Chapter 1
The History of Chinese Oil Industry Development

This book provides a comprehensive and detailed overview of the Chinese oil industry, its four historical phases and its role in the industrialization of China. Resources and exploration, pipeline development, refining and marketing, petroleum and natural gas pricing policies, and international cooperation are all addressed, as are conservation, renewable energy resources, and environmental impacts. Also presented are peak oil models and a forecast of future trends in fossil fuel supplies through the middle of the twenty-first century. A key consideration is the declining trend in energy returned on energy invested (EROI) for oil and natural gas in China, which technology seems unable to reverse and which will have enormous consequences for the country's economic prospects. The future of China's coal production is also discussed.

Despite discovering and producing oil for nearly two thousand years, China did not form an oil industry in its long history due to its feudal society that focused on farming products. Even at the beginning of the foundation of modern China in 1949, it was still an oil-poor country. How did China develop such a large oil industry in only 63 years? The next section examines that process.

1.1 Brief History of Oil in Ancient China

1.1.1 Discovery Records of Fossil Fuels in China

Chinese people had already begun to discover and utilize oil and gas more than two thousand years ago. They discovered natural gas first because natural gas has a greater tendency to be released from the geological stratum than oil, turning into flame when encountering wild fire or lightning.

Ancient people in China generally had fear of glaring lights from burning oil before they became familiar with the various uses of oil. They used to call it "Dragon Ointment." The ancient Chinese "Book of Changes" describes the scene

L. Feng et al., *The Chinese Oil Industry: History and Future*,
SpringerBriefs in Energy, DOI 10.1007/978-1-4419-9410-3_1,
© Lianyong Feng, Yan Hu, Charles A.S. Hall, Jianliang Wang 2013

in the Zhou Dynasty, dating back to as far as 3,000 years ago, of natural gas seepage on the surface of lakes and swamps (Yang and Zhang 1997). "The Classic of Mountains and Seas," compiled in 200 B.C., also describes the phenomenon of natural gas combustion after blowing out from the ground (Zhang 2002a).

In the late Han Dynasty, dated 1,900 years ago, local people in Yan'an, Shaanxi Province, had started to use oil for fuel and lubrication purposes. In the Jin Dynasty, Hua Zhang compiled the "Records of Natural Science" in 267 A.D., giving a detailed description of the characteristics of oil in Yumen, Gansu Province. In 561 A.D., people discovered oil in Xinjiang Province. They saw it as a treasure and made it a tribute to the court. In the Tang Dynasty, asphalt was discovered in Kuche, Xinjiang Province. The first person in Chinese history that formally proposed the name "石油(oil)" was Kuo Shen, the famous savant of the Song Dynasty. In his great work "Dream Pool Essays" (1086–1093), he made the forecast that oil "will be widely used in the future" (Bai 2009).

1.1.2   Early Application of Fossil Fuels in China

Natural gas and oil started to have wider application in every aspect of ancient Chinese society (Wang 2010a). For example, in 587 A.D., the guarding army in the besieged city of Jiuquan dispelled their enemy's attack by destroying the enemy's siege tools with the aid of burning oil. In the Song Dynasty, the central government set up the "Fierce Oil Workshop," in which thousands of people worked, specializing in production of the "oil weapon." In the Yuan Dynasty (1271–1368), oil was used for medical purposes and was applied to cure skin diseases of livestock. In the Ming Dynasty (1368–1644), people learned to extract and purify oil into lamp fuel (kerosene). The refining of oil also began early in China. The "Fierce Flaming Oil Mill," another name for the above-mentioned "Fierce Oil Workshop" is one of the earliest oil refining factories in the world. It began to produce oil products after rough processing. During 512 A.D. and 518 A.D., in the late North Wei Dynasty, Daoyuan Li introduced the process of refining oil into lubricants in his famous work "Commentary on the Waterways Classic." Thus in the long Chinese history of oil and gas discoveries and utilization, Chinese people, who were largely cut off from the rest of the world, had to rely on their own diligence and wisdom to invent many epoch-making techniques to use oil and natural gas.

1.1.3   Development of Drilling Technology in China

In China, natural gas wells were developed much earlier than oil wells. The reason is that natural gas was used to cook salt in Sichuan Province. This propelled the development of drilling technologies in China. Between 256 B.C. and 251 B.C. the first gas well for salt processing was drilled in Sichuan Province (Wang 1989). In the first Century A.D., natural gas was discovered in Linqiong, Sichuan province, and their manually drilled well of more than 100 m (330 ft) was, almost certainly, the deepest well in the world at that time.

Chinese people appear to have been the first to drill for oil. In the Warring States Period, some 2,200 years ago, the Chinese started to drill deep oil wells. The "cable tool drilling" technique, where a sharp stick was raised and dropped repeatedly on the end of a cable, was invented in the Song Dynasty (960–1279 A.D.). There are some indications that the Chinese had even used a primitive rotary bit. After the middle Ming Dynasty (1368–1644 A.D.), the "cable tool drilling" technique was greatly enhanced. As time passed, the drilling process was formalized and new materials were developed to guide and reinforce the casing strings while protecting well walls. The capability to prevent accidents was improved and new instruments were invented for fishing out apparatuses dropped into the well hole. During the period of Emperor Daoguang (1821–1850 A.D.), the cable tool drilling technique reached perfection and started to be divided into different work types, while well logging was beginning to be used in the drilling process. In 1850, the deepest well in the world called "Mezi," was drilled to a depth of 1,200 m. The Mezi oil well had a strong blowout. The explosion of crude oil into the air could be seen from 30 km away. In this well gas production was 50 thousand cubic meters per day, and by that time, is estimated that the annual natural gas production of artesian well fields in China had reached 100 million cubic meters. The improved cable drilling rigs and associated drilling techniques were basically the same as modern cable drilling in many respects, including: the main structure design and the working principles, the drilling tools, the down-hole accident treatment, the well logging and the casing. The main difference lies in the motive power, which was provided by humans and livestock. Daniel Yergin (1992) wrote in *The Prize* about the importance of Chinese technology "Around 1830, Chinese drilling techniques were introduced into Europe and the Americas and imitated afterwards." For example, the Drake Well in Pennsylvania of the United State, was drilled using steam-powered cable drilling techniques, most likely based on Chinese designs. Thus until around 1850, it appears that the Chinese were world leaders in drilling technology, but that changed when Western people successfully combined Chinese drilling techniques with steam engines to develop their wells. The result was that in the West, oil was exploited in very large quantities, while Chinese oil techniques fell behind.

1.2   The First Stage of the Chinese Oil Industry (1877–1949)

In the second half of the nineteenth century, the Opium War and the subsequent Western dominance, along with the intense oil exploitation in America, Russia, and other areas with oil production surpluses, opened the gates of China to Western oil products. This began with the purchase of kerosene, called "imported oil" at that time

(Lv 1983). The increase in imported oil sales caused a long-term trade deficit in China's foreign trade, which eventually gave birth to the modern Chinese oil industry.

Modern oil wells in China appeared first in Miaoli, Taiwan (Zhang 2004). In 1877, the Qing government bought a new steam-powered cable drilling machine from the United States, and hired two American technicians to drill in Taiwan. Oil was found in the spring of 1878, and the crude oil production was 0.75 tonnes per day afterwards (there are about 7 barrels in a tonne of oil, depending on type per tonne: most of the data reported here is from China Energy Statistic Yearbook (CESY) 2009 and other years). This first modern oil well in Taiwan was the prelude to the Chinese modern oil industry.

In 1907, in Yanchang, Shaanxi, the province with the earliest record of oil seepage, people began to use modern drilling machines to extract oil. They hired Japanese technicians to successfully drill their first industrial well. Oil was found 81 m (260 ft) below the surface, and production yield reached 1 tonne per day, giving nationwide recognition to the project. In 1912, after the Japanese technicians returned to their country, the 12th well in Yanchang was drilled. This was the first well drilled entirely by Chinese workers using modern drilling technology (Liu and Ma 2006). In 1909, in Xinjiang Province, a small refining factory was bought from Russia and installed in Dihua, establishing the Urumchi factory. In addition, cable drilling equipment was bought and oil was obtained under the guidance of Russian technicians (Deng 1992). These incidents were the beginning of the modern oil industry in Xinjiang Province.

In February of 1914, the Chinese government signed a contract with the Mobil oil corporation to explore and exploit oil in Shaanxi Province. However, in 1917, this 3-year Sino-American cooperation scheme turned out to be a failure: the Mobil geologists had obtained negative results from the oil exploration conducted in Shaanxi Province. There even was a prevailing saying among the international community, "there is no possibility of large-scale oil production whatever in marine or terrestrial strata in China." Thus China was classified as one of the "oil-poor countries." The most representative figure promoting this view was American geologist and Stanford University professor, Blake Velde who wrote the book "The Oil Resources in China and Siberia" (Song 2005).

Through the long period of more than seven decades (1877–1949) the Chinese oil industry had discovered just a few small oilfields and had fallen far behind the rest of the world (Zhang 2001). Although oil was discovered in the provinces of Shaanxi and Xinjiang, the quantity was small. The exception to this trend may be the Laojunmiao oilfield, which was reputed as "China's first mine in modern oil industry." Most Chinese historians consider that Laojunmiao symbolized the commencement of Chinese oil industry in the true sense: it was the largest and most modern oilfield in China in the1940s, making special contributions to the War effort and the liberation of China from Japanese dominance (Jiang and Shi 2009). However, the cumulative proven oil reserves had reached only 70 thousand tonnes by 1949.

1.3   The Second Stage of the Chinese Oil Industry (1949–1978)

In the early stages of the People's Republic of China (PRC), the oil industry recovered and oil production started to grow again. Nevertheless, the reputation of China as an "oil-poor country" was not thrown away. From 1949 to 1959, oil exploration in China was concentrated in the western regions, mainly exploring at the edges of outcrop areas and shallow tectonneic areas in Jiuquan, Junggar Basin, Tarim Basin, Qaidam Basin and Ordos Basin. In 1952, the crude oil output for China was only 4.35 million tonnes, which could meet only one quarter of national needs. In this stage, Yumen city, located northwest of Gansu Province, was regarded as the most promising region for quickening the pace of domestic oil production (Wang 2009a).

From 1950 to 1952, along with increased exploration efficiency, oil production wells increased to 30, crude oil production yields increased to 375 thousand tonnes and oil refining capability was greatly enhanced. In 1953, China began the first "five-year plan," in which the Yumen oilfield was listed as one of the key construction projects. During this period, with the aid of experts from the Soviet Union, there was a surge in oil production of the Yumen oilfield. On October 8, 1957, the first oil industrial base of the PRC was established in Yumen. Production yields reached 755 thousand tonnes that year, accounting for 87.8% of oil production nationwide. In 1958, oil production of the Yumen oilfield reached 1 million tonnes (about 7 million barrels), and peaked at 1.4 million tonnes in 1989 (Li 2010).

At the same time, there was a substantial growth of oil refining in many synthetic oil plants in three northeastern provinces, mainly through the expansion and rebuilding of existing infrastructure. In the meantime, a large shale oil plant was constructed in Maoming, Guangdong. In 1959, production yields of artificial oil for the entire country reached 970 thousand tonnes. The oil refinement industry that started on a very weak base had developed along with the expansion of eight oil refineries with an annual processing capability between 100 and 1,000 thousand tonnes of oil located in Shanghai, Kara may, Lenghu, Lanzhou and Dalian. In 1959, the production of gas, coal, diesel and lubricants was around 2.35 billion tonnes in China. During these years, China had achieved 42.6% self-sufficiency of primary oil products (Wang et al. 1960).

In accordance with the first "five-year plan," oil exploration began in the northwestern region in China. On October 29, 1955, well No.1 in Heiyou Mountain hit oil. Located in the northwest side of Junggar Basin, Xinjiang Province, this oil well marked the discovery of the first large oilfield in what is called now the Karamay oilfield play. After a few years of construction, prior to the discovery of the Daqing oilfield, the Karamay oilfield became the largest oil production base in China, with an annual oil production of 330 thousand tonnes and reaching 1.63 million tonnes in 1960 (Hu and Wang 2011). In the meantime, news of oil discovery came out from Qaidam Basin in Qinghai Province and Central Sichuan Province. The experience of these oilfields provided invaluable knowledge for the later very successful exploitation of oil in China.

In 1958, the government of China shifted its main oil exploration efforts to the eastern region. This critical decision turned over a new page for the development of the Chinese oil industry. On September 26, 1959, oil flowed from well Songji No.3 in the Songliao Basin marking the birth of Daqing oil field, which had a profound impact on the history of the Chinese oil industry. In the 3 years that followed, people from all walks of life all over China migrated to Daqing due to the economic boom, and an unprecedented enthusiasm for oil began. In 1960, Daqing's proved oil reserve was 400 million tonnes and the first carload of crude oil was produced. By the end of 1963, Daqing's crude oil production capability reached 5 million tonnes a year and its cumulative production was 15 million tonnes. The successful efforts in Daqing were a giant leap in the history of Chinese oil industry development and changed its outlook fundamentally. On December 3, 1963, Premier Zhou declared solemnly, "China's oil is basically self-sufficient and the epoch of Chinese people using imported oil will be gone forever" (Kang 2001).

Around the period of oil development in Daqing, exciting news came frequently from the Chinese oil exploration front as new oilfields were discovered successively. In September of 1959, the Jilin oilfield was discovered by the well Fu No. 27; in 1961, the Shengli oilfield by well Hua No.8; in 1963, the Dagang oilfield by well Huang No.3; in 1965, the Liaohe oilfield by well Liao No.2, the Jianghan oilfield by well Wang No.2. Subsequently, oil efforts were launched in Sichuan, Jianghan, Changqing, Liaohe, Jilin, Jiangsu, Henan, Renqiu, Dongpu and other provinces, all of which achieved good oil discoveries, although none as important as Daqing. In the meantime, a group of oil refinement plants were gradually built around China.

In 1965, the goal of overall self-provision of oil products was realized in China. From 1966 to 1976, China went through the "Cultural Revolution," during which the oil industry, like other industries, suffered serious interferences. However, oil workers in Daqing oilfield overcame all sorts of difficulties and continue producing oil. In this period, oil production in the Daqing oilfield reached 10 and later 50 million tonnes in 1976. In addition, oil drilling platforms were constructed offshore. Oil pipelines and refineries were also constructed around China. In 1970, Changqing oilfield was discovered by well Qing No.1, and Henan oilfield by well Nan No. 5. In 1975, The Renqiu oilfield was discovered by well Ren No.4. In 1978, the crude oil production yields of China reached 104 million tonnes (Chu and An 2004), with the Daqing oilfield accounting for about half of the production. The annual production of Shengli, Dagang, Xinjiang and Jilin oilfields were 19.46 million tonnes, 3 million tonnes, 3.53 million tonnes and 1.85 million tonnes respectively. The annual natural gas production was 13.73 billion cubic meters. By that time, China had made its way from an "oil-poor nation" to the rank of the eighth largest oil producer in the world.

From 1966 to 1978, Chinese crude oil production increased at an annual rate of 18.6%. During the same period, the oil industry provided the necessary energy for the development of the Chinese national economy and became one of the primary sources of foreign exchange for China. In the last 40 years, Daqing's crude oil exports alone have accounted for 52 million tonnes of oil.

1.4   The Third Stage of the Chinese Oil Industry (1978–1998)

The year 1978 symbolizes one of the landmark turning points in Chinese history. The Third Plenary Session of the 11th Central Committee of the Communist Party of China was held that year and the program of "Reform and Opening up" was initiated, formalizing the opening of Chinese trade with other countries. The Chinese oil industry also stepped into a new developmental stage in this context.

During these 20 years, the efforts of the Chinese oil industry included efforts to expand the scope of autonneomy rights of oil companies (this is similar to the Western concept of "decentralization," although the central government in China still maintained considerable control), implementing responsibility management contracts, insisting on the policy of "Bring in, Going Global" which opened up new rounds of talks in oil-resource related fields, reforming the oil administrative system and building up national oil companies while taking part in international oil and natural gas co-operations. The two main landmark events for the oil industry were the abolishment of the State Ministry of Oil Industry in 1988, and the restructuring of three major Chinese oil companies in 1998.

After crude oil production exceeded 100 million tonnes in 1978, reserve growth could not keep up with production growth due to long-term funding insufficiency, reduced exploration workload and also the accelerated decline in the yield of old oilfields. As a pillar industry of the Chinese economy and a vital source for earning foreign exchange, maintaining stable oil production of 100 million tonnes became an urgent task for the Chinese oil industry. Consequently, in 1981, the government put forward the policy of "100 million tonnes crude oil production contract" (Jiang et al. 2008). The comprehensive implementation of this contract provoked a series of profound reforms with regard to the internal mechanisms of the oil industry, accelerating the development of the Chinese oil industry.

For example, the production yields of the Shengli oilfield reached 19.46 billion tonnes in 1978, decreased to 17.59 million tonnes in 1980, and then showed signs of further decline. However, after the "All-round Contract," Shengli oilfield's production yields maintained 3 years of consecutive growth.

The program for "100 million tonnes production contract" gave results that were more powerful than expected. Oil exploration emerged from the difficult situation of insufficient funds and with effective implementation of the "bring in" policy, the Chinese oil industry stepped on the route for rapid development. In 1983, the oil ministry signed a contract for exploratory development services with 10 seismic crews from the United States and France, whose completely different working methods, management style, and effective techniques broadened the horizon of the Chinese oil industry. This was a very audacious effort for the early stage of the Reform and Opening Up program in China.

By 1985, a total exploration fund of 4.29 billion dollars had been raised for Chinese oil exploration (Zhang 2000). Chinese seismic technology for oil exploration had begun a technical change to digital seismology and electronically computerized processing, ultimately reaching the top level of international petroleum exploration.

Afterwards, new bit manufacturing techniques were introduced from the Houstonne Corporation in the United States, and the Jianghan Bits Factory was built. These improvements greatly enhanced the technological and managerial level of Chinese oil equipment and ended the stagnation period of the Chinese oil industry, starting the period of "benign development." By the mid- 1980s, through the hard work and practice of numerous geological workers, there were major improvements in the Chinese oil geological theory. This gradually evolved into key petroleum geological theories, which have served as guidance towards the healthy development of the Chinese oil industry.

In China, the "100 million tonnes production contract" was the largest industry contract in the 1980s. It raised the enthusiasm of oil companies enormously and relieved the stress of funding shortages, promoted the stable growth of oil production, raised the level of oil workers' living standards, and increased crude oil production to 125 million tonnes in 1985 (Chen 2008). Along with the change of overall atmosphere in the Chinese oil industry, there was a radical change of its principle structure. The former administrative government institution was transformed into three oil companies. In 1982, the China National Offshore Oil Corporation (CNOOC) was founded; the next year, the China National Petroleum & Chemical Corporation (Sinopec) was founded, taking charge of 39 vital oil-refining, petrochemical and chemical fiber enterprises; in 1988, the Ministry of Petroleum Industry was canceled and the CNPC was established to manage exploration, development and production of oil and natural gas resources onshore. The founding of three national oil companies was a successful and vital element in the reform of the Chinese structure for industrialization management in the early stage of "Reform and Opening Up." After a period of time, these three national oil companies successfully promoted the development of the Chinese oil industry. However, problems of separation between upstream and downstream links in the chain of the petroleum industry, between domestic and foreign trade, and between terrestrial and marine areas appeared as cause for concerns, with the contradiction between unbalanced development and uneven distribution.

In 1998, the Chinese government implemented the restructuring of their petroleum and petrochemical strategies, changing the administrative allocation and swapping assets by reorganizing the former National Petroleum Corporation and National Petrochemical Corporation into two large-scale petroleum and petrochemical companies. These companies achieved the integrated operation of the industry including the upstream and downstream chains, the upper and lower components of manufacturing and production-supply-marketing, and of domestic and foreign trade. On the one hand, the restructuring optimized the industrial structure of the oil industry, and on the other it adjusted benefit-based relationships which served to incentivize the different parts of the industry and shaped the pattern of competition and cooperation. These actions provided a solid base for the internal restructuring and listing of the two large companies. Shortly afterwards, the CNPC, Sinopec, and the CNOOC launched internal corporate restructuring and formed their respective stock companies, all of which were successfully listed in overseas markets. This marks a historical breakthrough in the property rights reform of state-owned oil companies and their successful debut in the international capital markets. As in much of the rest of the Chinese society, these actions showed how China could achieve a unique economic structure of somewhat private firms, which are ultimately under the directorship of the Chinese national government.

1.5 Vital Oil Fields in the History of the Chinese Oil Industry

Chinese oil and gas resources are located mainly in Bohai Bay, Songliao Basin, Tarim Basin, Ordos Basin, Junggar Basin, Sichuang Basin, and Qaidam Basin. The main oilfields in China are the Daqing oilfield, the Changqing oilfield, and the Tarim oilfield. In order to present a general perspective of oil reserves and production yields in recent years, we will highlight 14 oil-gas fields and their respective exploration and production statuses. These include the above-mentioned oilfields plus the Shengli oilfield, Yanchang oilfield, Xinjiang oil field, Tahe oilfield, Liaohe oilfield, Jilin oilfield, Dagang oilfield, Bohai oilfield group, Huabei oilfield, Zhongyuan oilfield and Southwest Oil &Gas fields. The major petroleum production areas are given in Fig. 1.1.

1.5.1 Daqing Oil Field

Daqing oilfield, China's largest, was discovered and began production in 1959. Production capacity reached 6 million tonnes in 1963 and actual crude oil production yield was 4.4 million tonnes in that year. In 1976, production reached a rough maximum of just over 50 million tonnes, half of the aggregated national target, and afterwards maintained a stable annual yield of over 50 million tonnes for the next consecutive 27 years. Its peak reached 56 million tonnes in 1997. However, in 2003 production fell to 48.4 million tonnes. There was a further decline in production afterwards, to 40 million tonnes in 2009. The average decline rate between 2003 and 2009 was 3.1%, a rate that is roughly characteristic of many large fields (Zhou 2009). The goal set by CNPC in 2000 was that Daqing should maintain stable oil production of at least 40 million tonnes for 10 years, which meant that the Daqing oil field would have had to maintain annual production yields of over 40 million tonnes during the next 3 years. This goal relied mainly on the technique of enhanced recovery, such as water injection. Exploration and exploitation in Daqing has entered into a mature or post mature stage and production will undoubtedly continue decreasing in the future. There are increasing difficulties in oil extraction and the water cut has already exceeded 90%. Therefore, it is a great challenge for the Daqing oilfield to achieve the goal of 40 million tonnes in the following years. The decline of this important oil field has come as a blow to many oil workers who believed that Chinese technology would be able to keep the oil flowing at the maximum rate.

Fig. 1.1 Distribution of vital Chinese oil fields. *Source*: EIA, 2011. U.S. Energy Information Administration

1.5.2 Shengli Oil Field

The Shengli oilfield, located in the Yellow River Delta area, along the Bohai Bay coastline in the northeast of Shandong Province, was discovered and began production in 1961. In 1973, crude oil production reached 10.8 million tonnes; in 1984, oil production exceeded 23 million tonnes and, 3 years later, it exceeded 30 million tonnes. In 1991, annual production peaked at 33.5 million tonnes, followed by a gradual decline. After 8 years of successive decline, i.e., after 1999, oil production maintained a steady pace and climbed slowly upwards until 2004. Oil production reached 27.9 million tonnes in 2009, ranking second in China (Yang 2010a, b).

The natural gas yields of the Shengli oilfield were relatively high from 1975 to 1998. The average annual production yield was 1.2 billion cubic meters, and peak production occurred in 1989 at about 1.5 bilion cubic meters, followed by a decline afterwards. Since the beginning of the twenty-first century, the decline in natural gas production of the Shengli oilfield was reversed and production started to climb slowly upwards for 4 years, after which it began to fall again. Total natural gas production as of 2008 was 770 million cubic meters.

1.5.3   Changqing Oil Field

The Changqing oilfield, whose main operation fields are located in the Ordos Basin, was discovered in 1971 and began producing in 1975, but its production level has been relatively low, fluctuating around 1–1.4 million tonnes. However the discovery of the Ansai oilfield and Jing'an oilfields, with respective reserves of over 200 million tonnes and over 300 million tonnes during the middle 1980s and late 1990s, allowed a slow climb upwards since 1991 (Hu and Wu 2007), to more than 2 million tonnes in 1995, followed by a rapid increase afterwards. In 2001, crude oil production exceeded 5 million tonnes, and in 2009 15.7 million tonnes. The average annual growth rate reached 14.8%. In 2010, oil production of the Changqing oilfield ranked third in the nation.

Natural gas production of the Changqing field grew rapidly in the twenty-first century, with an annual growth rate of 29.3%. Production reached 19 billion cubic meters in 2009, the first in the nation, giving it the title of "The Biggest Gas Producing Region." The total yield of natural gas and oil exceeded 30 million tonnes of oil equivalents, ranking second in the nation.

1.5.4   Yanchang Oil Field

Yanchang oil field, which is located in north Shaanxi Province, is a low permeability reservoir. In 2005, it was moved from public ownership to the Shaanxi Yanchang Petroleum (Group) Corp. Ltd. (abbreviated as Yanchang Petroleum Group), following restructure (Yanchan oil field overview website). A main difference from the above-mentioned oil fields is that Yanchang Petroleum Group is not owned by the state. Instead, its share is controlled by Shaanxi government and is classified as a local enterprise. In recent years, Yanchang Petroleum Group has been launching oil exploration in Yanchang oil fields, and expanded resources area for exploration by 2,000 square meters in Wuqi, Dingbian, Jingbian, Fuxian, Ansai, Zhidan, and other regions. New breakthroughs occurred in deep exploration and demonstrated reserves increased by 95 million tonnes. Exploration teams outside Shaanxi province discovered oil and gas wells and have produced low yield test wells in Inner Mongolia. In 2008, crude oil production reached 10.9 million tonnes. In 2009, reserves were 310 thousand tonnes, and production reached 11.2 million tonnes, ranking fifth in the nation. In spite of unremitting efforts and good results achieved in recent years, Yanchang oil field was faced with the difficulty of limited controlled resources and a sharp production decline in each the field's wells.

1.5.5   Xinjiang Oil Field

Xinjiang oil field was discovered and began production in 1995. It is located in Karamay, Xinjiang Province. Its main exploitation area is in the Junggar Basin, and the rest is in the Tarim Basin. Production yield of the Xinjiang oil field has maintained a steady and rapid growth, surpassing 3 million tonnes in 1988, 9 million tonnes in 2000, and 12.2 million tonnes in 2007. In 2009, crude production was 10.9 million tonnes, ranking sixth in the nation. In 2009, it had a 10.8% decline from 2008, the first decline in recent years (Wang 2010a).

In 2006, the Mahe natural gas field was discovered in the Xinjiang oil field, with reserves of 30 billion cubic meters. In December, 2008, the first stage of the Kelameili Gas Field construction project was completed. This field, located in Junggar Basin, has a reserve of more than 100 billion cubic meters. Its natural gas yield per day surpassed 10 million cubic meters. By the end of 2009, forecasted reserves of natural gas in Xinjiang reached 53 billion cubic meters. Similar to oil production, natural gas yield maintained growth, especially after 1990. In 2009, natural gas yields reached 3.6 billion cubic meters, 200 million cubic meters more than the previous year. Apparently, there is a huge potential for future growth.

1.5.6   Tarim Oil Field

The Tarim oil field, located in the Tarim Basin in southern Xinjiang Province, was discovered and began production in 1989. Crude oil output increased year by year until it peaked in 2008 at 6.5 million tonnes. In 2009, its production of 5.5 million tonnes ranked tenth in the nation. Natural gas production increased rapidly from 1.35 billion cubic meters in 2004 to 18.1 billion cubic meters in 2009. The annual growth rate is 67.9%, and gas production yield ranked second in the nation in 2009. The sum of oil and gas production which is 20.7 million tonnes ranked fourth in the nation. Afterwards, the production level was successfully maintained above 20 million tonnes (Song and Jiang 2008).

The Tarim oil field provides one of the main sources of gas for the "West to East Gas Transmission" project. In 2009, oil and natural gas reserves reached the maximum for the field. The completion of the large chemical fertilizer unit in the Dina No. 2 Condensate Field was a key project for Tarim Petrochemical Corporation. Production capability is still building with the development and testing project of the Central Tarim No.1 Gas Field. In 2010, CNPC continued its exploration and exploitation efforts of the Tarim oil field, of which one of the key investment areas is the Central Tarim No.1 Gas Field, located in the Hinterland of the Taklimakan Desert, also called "Sea of Death." Thus the Tarim oil field is still expanding in production and chemical utilization.

1.5.7   Tahe Oil Field

The Tahe oil field is a subsidiary of the Northwest Oil field Branch Company of Sinopec. Located in the north Tarim Basin, Tahe oil field was discovered and began production in 1997, followed by an increase in production afterwards, leaping from the initial 390 thousand tonnes to 6.6 million tonnes in 2009, ranking the eighth in the nation. Natural gas yield reached 1.3 billion cubic meters. Due to the depth and the characteristics of the oil in the oil field (heavy with high contents of hydrogen sulfide), Tahe employed 13 multi-well units using water injection, which resulted in an effective ratio (which means wells that work over total wells) of over 80%. In 2009, in view of the situation of a rapidly increasing water incursion into the main block, the Northwest Oil Field Branch Company took active measures to improve the oil field operations, such as controlling waterflooding, stabilizing oil production and other comprehensive managerial measures. The overall effectiveness ratio of the above-mentioned measures is 85.7%, up by 17.4% on a year-over-year basis (Kang 2003). There is still potential for the development of the Tarim oil field in the future.

1.5.8   Liaohe Oil Field

The Liaohe oil field, located in the middle and downstream areas of the Liaohe River, East Inner Mongolia and the shallow sea areas of Liaodong Bay, was discovered in 1958 and began production in 1970, followed by a gradual increase in production yield afterwards. Its peak production was reached at 15.5 million tonnes in 1995, ranking it in third place in the nation at the time (Cui 2000). Afterwards, production declined. In 2009, production had fallen to 10.2 million tonnes, making it seventh in the nation. As the nation's largest production base of heavy oil and high solidification point oil, the Liaohe oil field has been continuously explored for 40 years, but its production has fell consecutively for the past 15 year. Since 1995, natural gas production continued along a decreasing trend, dropping to 810 million cubic meters in 2009.

1.5.9   Jilin Oil Field

Jilin oil field is located in Fuyu, Jilin Province. Oil and natural gas exploration and exploitation was conducted in two major basins in Jilin Province. Production began in 1960, with fluctuating growth. Crude oil production in 2008 was 6.6 million tonnes, a record in its history. In 2009, its total production of 6.1 million tonnes ranked ninth in the nation, even after a small decline. In that year, crude oil production of the Jilin oil field decreased by 500 thousand tonnes, while natural gas production increased by 620 million cubic meters on a year-on-year basis, reaching 1.2 thousand cubic meters. Thus, it maintained the oil and gas equivalent production at the 7 million tonnes level. Moreover, in recent years, the geological reserve of natural gas in Jilin oil field exceeded 0.2 trillion cubic meters, creating a bright exploration outlook for natural gas exploitation in Jilin. On December 25th, 2009, Changling gas field, located in south Songliao Basin, was built and put into production. The annual natural gas production capability reached 1 billion cubic meters (A R.H 2010).

1.5.10 Dagang Oil Field

Dagang oil field, located in Dagang region, Tianjin, comprises Dagang region and Yourdusi Basin in Xinjiang Province. Before 1975, there was a stable growth in oil production. Peak production of 4.5 million tonnes was reached in 1975, followed by a slow rise with large fluctuations. In 2008, its oil production reached 5.1 million tonnes. In the aspect of natural gas production, peak production was reached at 927 million cubic meters in 1979. During the period that followed, until 2006, production fell gradually. Total production in 2006 was 325 million cubic meters. Production began to increase, and production yields in 2007, 2008 and 2009 were 542 million, 554 million and 540 million cubic meters respectively (CNPC website 2010).

1.5.11 Bohai Oil Field

In recent years, the discovery of many oil and natural gas fields in the Bohai sea area has laid the foundation for Bohai oil field. Bozhong well 2-1 and Qinhuangdao well 29-2 were discovered through two successfully drilled medium-sized oil fields. High yielding oil and natural gas flows were obtained in both discovery wells. In Jinzhou's 20-2 north structure, which is northeast of 20-2 gas field, Liaodong Bay, the Jinzhou prospecting well 20-2N-1 made a new discovery. The well was tested to flow at an average rate of 1.9 thousand barrels of oil and 123 million cubic meters of natural gas per day. As China's largest oil base on the sea, the Bohai oil field displayed stable growth in production in recent years. In 2008, the total oil production was about 11 million tonnes, and in 2009, about 13.5 million tonnes, ranking fourth in the nation. There remains much space for further development in this oil field (Chang 2005).

1.5.12 Huabei Oil Field

Huabei oil field, located in Renqiu city in Jizhong plain in central Hebei Province, comprises oil producing areas distributed in Beijing, Hebei, Shanxi and Inner Mongolia. Built and put into production in 1975, Huabei oil field's peak production was reached at 17.3 million tonnes in 1979, followed by a gradual decrease afterwards. In 2008, production fell to the level of 4.4 million tonnes. With respect to natural gas production, growth was maintained until 2004, and the total production yield in 2004 was 585 million cubic meters, followed by a gradual decline and then steady production periods afterwards. Gas production in 2008 was 553 million cubic meters, and in 2009, 550 million cubic meters (Su et al. 2010).

1.5.13   Zhongyuan Oil Field

Zhongyuan oil field is located in Puyang, Henan Province. It was discovered in 1975 and began production in 1976. Peak production was reached in 1988 at about 7.2 million tonnes, followed by a continuous decline (Cui 2004). In 2008, production had dropped to 3 million tonnes. Peak production of natural gas was reached in 2004 at 1.8 billion cubic meters. But in 2009, natural gas yield fell to the level of 800 million cubic meters, and the rate of gas production has been declining year by year.

1.5.14   Southwest Oil and Natural Gas Field

Southwest Oil and Natural Gas Field, owned by CNPC, is one of the earliest gas fields explored in China. Gas production started in 1950. It is mainly distributed in Sichuan Basin and Xichang Basin, and has become one of the primary natural gas production areas. Natural gas production has increased regularly since 1950, especially after 2000 and especially 2004, when production surged from 9.8 billion cubic meters to 15 billion cubic meters in 2009 (Wang 2010b).

Chapter 2
Developmental Features of the Chinese Petroleum Industry in Recent Years

China's oil industry has developed for more than 60 years since the establishment of China in 1949, creating many achievements in exploration, development and production to meet the increasing needs for oil. What is the current situation of the Chinese oil industry? The next section analysis that.

2.1 Exploration and Development of Petroleum in China

2.1.1 The Progress of Exploration and Development

In 1998, the Chinese government decided to restructure the petroleum and petrochemical industry in accordance with the principles of upstream and downstream integration. CNPC and Sinopec were officially established. Later, PetroChina Company Limited merged with CNPC, China Petroleum & Chemical Corporation attached to Sinopec, and CNOOC Limited merged with CNOOC, were founded successively. That same year, China pursued a decade-long program of intensive petroleum exploration. During this period, four exploration fields including eastern, western, marine and overseas regions were launched (Li 2009). Numerous exploration accomplishments were achieved with this new policy. Oil reserves and output steadily increased on the basis of this high level effort, despite even greater difficulty in exploration. In the eastern area, constantly increasing research on oil and gas geology and accumulation, and continuous improvement of techniques and methods for exploration, made the oil exploration more effective. Proven oil geological reserves increased annually to a high level of 100 million tons. In the western and marine areas, discovery of the giant Tahe oilfield in Tarim Basin, the breakthrough of giant Xifeng oilfield in Ordos Basin and the demonstration of Penglai 19-3 and other oilfields in Bohai Sea, increased the reserves in these regions. Overseas, China won the bid for exploration and development project of

L. Feng et al., *The Chinese Oil Industry: History and Future*,
SpringerBriefs in Energy, DOI 10.1007/978-1-4419-9410-3_2,
© Lianyong Feng, Yan Hu, Charles A.S. Hall, Jianliang Wang 2013

Block 1/2/4 in Sudan; subsequently, a number of large and medium-sized oilfields were discovered after the contract had been signed in 1997.

Meanwhile, with the restructuring of the petroleum industry, local companies were able to undertake exploitation and production in accordance with the new law, some attaining rapid development. As a result, a new management pattern of oilfield exploitation took shape. In this petroleum stage, the main problems in China were as follows: many developed oilfields entered into a stage of high water-cut and high levels of exploitation with declining levels in production; the shortage in petroleum resources was still the main restricting factor for the development of many oilfields; the drilling situations became more and more complicated, making the existing technology inadaptable, finally, reconstruction turned out to be arduous because of the aging and severely corrosive ground systems of old oil fields. Based on the existing problems and difficulties, technical countermeasures were taken as follows: improving the technology of secondary oil production according to the characteristics of each oil field; developing and perfecting the supporting technology and industrial application of tertiary recovery; promoting the supporting technology in the economic and efficient development of low permeable oil fields; improving the effectiveness of thermal recovery of heavy oil and developing the succeeding technology of steam and soaking in the later periods of heavy oil; and developing new technologies for evaluating and decision-making in the exploitation of oilfields. Several other improvements for the petroleum industry in China included: accelerating the progress of integration of exploration and development; making efforts to control the scale of investment to reduce production costs; improving the yield and productivity; taking measures to increase production by adjusting and optimizing the physical and management system; changing scientific management methods; and increasing technology investment to make the size of the enterprise match the needs of development, improving reservoir management, and enhancing the integral level of oilfield development (Han 2010).

2.1.2 Evaluation of Oil and Gas Resources in China

According to the "Bulletin of Chinese Land and Resources in 2009" (MLRPRC 2010) issued by China's Ministry of Land and Resources (MLR), in 2009, proven geological reserves of oil increased by 1.1 billion tons—the fourth straight year that new proven geological reserves exceeded 1 billion tons. Newly proved reserves of natural gas were 723 billion cubic meters, hitting a historically high level. During the same period, three large oil companies made remarkable exploration achievements. The construction of western oilfields was on the rise. A series of oil and gas discoveries were made in Bohai Bay. CNPC continued to implement the project of peak reserves growth and achieved a series of strategic discoveries and significant breakthroughs in major exploration regions, such as Qaidam basin, Erdos basin, Tarim basin, Junggar basin, Hailaer-Talmud Taga basin and Sichuan

Table 2.1 Petroleum resources of three main blocks of China

	Eastern		Western		Offshore	
	Total (million tons)	Share (%)	Total (million tons)	Share (%)	Total (million tons)	Share (%)
Prospective oil	41.8	38.5	27.1	29.9	15.2	14.0
Geological oil	32.4	42.4	17.5	22.9	10.7	14.0
Recoverable oil	10.0	47.3	4.8	22.6	2.9	13.8

Table 2.2 Petroleum resources statistics of main petroliferous basins (million tons)

Basin name	Prospective oil	Geological oil	Recoverable oil
Bohai bay	27.5	22.5	5.5
Songliao	14.4	11.3	4.6
Tarim	11.4	8.1	2.4
Erdos	8.8	7.4	1.7
Junggar	8.5	5.3	1.3
Qiangtang	8.5	5.1	1.1
Pearl River Estuary	2.9	2.2	0.8
Qaidam	1.5	1.3	0.3
Cuoqin	2.2	1.1	0.2

basin. Sinopic increased exploration investment and strengthened its trap reserves. Sinopec made new exploration breakthroughs in the Leikoupo group in the northeastern part of Sichuan province. It also made significant progress in exploration in Xinjiang Tahe-Tuoputai region, and exploration discoveries in new strata series in the Eastern old region and Western new blocks. CNOOC made 15 self-supported oil and gas new discoveries in Chinese seas, and evaluated 11 oil-gas-bearing structures independently and successfully. In other words, in 2009, oil and gas reserves in China hit record highs, while resource development realized the goal of "stable output of oil and increased yield of gas."

In China, a number of nationwide petroleum resource evaluations were launched respectively in 1994, 2002 and 2005. There are a total of 418 sedimentary basins in the entire nation. By 2005, more than 150 sedimentary basins have been selected for oil and gas exploration. Among them, 24 basins were discovered with industrial oil-gas fluid and 34 were found to have evidences of oil and gas. As the national assessment of oil resources in 2005 revealed (MLR 2005), most oil resources are distributed in the Eastern, Western and Offshore regions of China. The evaluation reported three major resource types, including *prospective sources* that refer to total estimated petroleum initially in place, *geological resources* that refer to discovered petroleum initially in place, and *recoverable resources* that mean ultimate recoverable resources. Together, their prospective, geological and recoverable oil accounts for 82.4%, 79.3% and 83.7% respectively of the nation's whole. The detailed results are presented in (Tables 2.1 and 2.2).

Table 2.3 Natural gas resources of three main blocks of China

	Central		Western		Offshore	
	Total (10^{12} m^3)	Rate (%)	Total (10^{12} m^3)	Rate (%)	Total (10^{12} m^3)	Rate (%)
Prospective gas	18.0	32.3	15.9	28.4	12.7	22.8
Geological gas	10.1	28.9	11.6	33.1	8.1	23.1
Recoverable gas	6.4	28.9	7.5	33.9	5.3	23.9

Table 2.4 Natural gas resources statistics of main gas-bearing basins (10^{12} m^3)

Basin name	Prospective gas	Geological gas	Recoverable gas
Tarim	11.3	8.9	5.9
Sichuan	7.2	5.4	3.4
Erdos	10.7	4.7	2.9
East China Sea	5.1	3.6	2.5
Qaidam	2.6	1.6	0.9
Songliao	1.8	1.4	0.8
Yinggehai	2.3	1.3	0.8
Qiongdongnan	1.9	1.1	0.7
Bohai Bay	2.2	1.1	0.6

Natural gas resources in China are mainly located in the central, western and offshore regions (Table 2.3). In the Tarim, Sichun, Erdos, Qaidam and five similar gas-bearing basins, whose geological resources are more than 1 trillion cubic meters (Table 2.4).

In the future, Zhou and Tang (2004) recommend that new data, new theories and new technologies should be applied to existing places such as Songliao, Bohai Bay, Ordos, Sichun, Junggar, Tarim, Qaidam and Tuha basin. Also, the greater efforts must be made for the evaluation of new regions and new basins in order to find new fields for further exploration. Moreover, they recommend that new formation mechanism of oil and gas accumulation, and new oil and gas distribution laws need to be developed. Geological and geochemical data obtained from oil and gas exploration should be made more publically available, and new resources assessment methods geared to international standards should be adopted.

2.2 Production and Consumption of Petroleum in China

China is one of the world's largest oil producers. In 2009, the oil output of China amounted to 4.9% of global production. From 1999 to 2009, the notable feature of China's oil production was a steady growth trend at about 1.8% per year (Fig. 2.1).

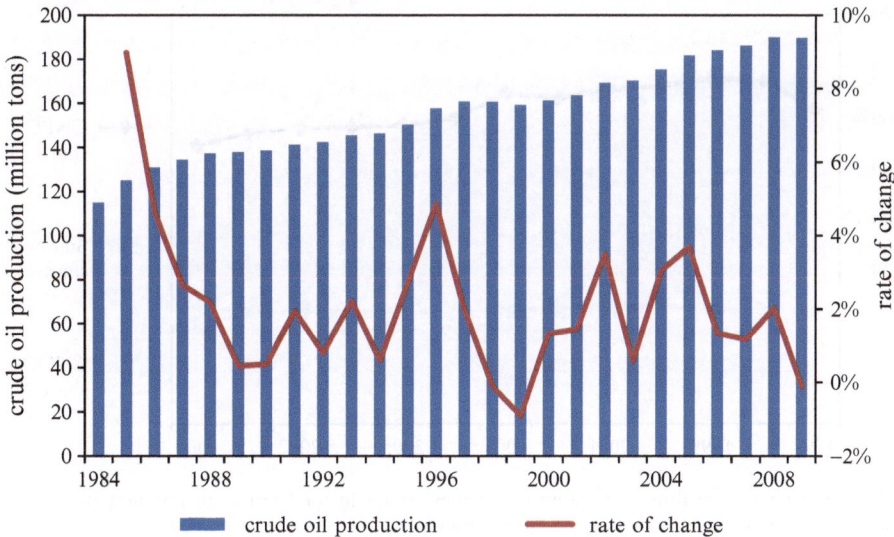

Fig. 2.1 China's crude oil production in 1984–2009. *Source*: International Petroleum Economics (Chinese journal). *Note*: Data is calculated in company range and has little differences with that from the Bureau of Statistics of China

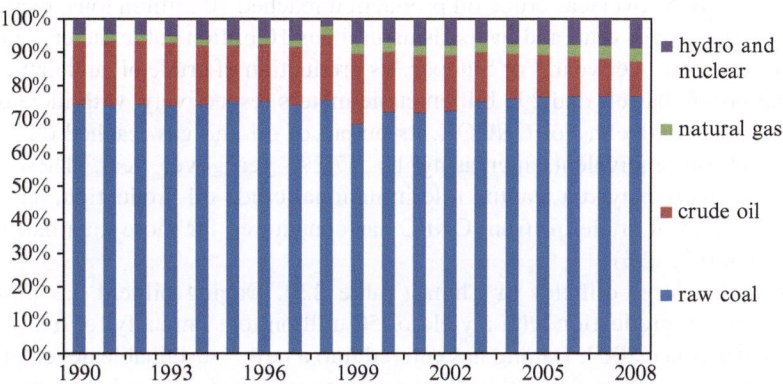

Fig. 2.2 Structure of energy production of China 1990–2008. *Source*: CSY 2010

However, there are some indications from recent years that this increase may not continue because of the increasing difficulty in getting the oil (see Chap. 4) (Fig. 2.2).

According to the NBSC, in 2009, total national crude production was 189 million tons with a 0.4% drop year over year and natural gas 83 billion cubic meters with a 7.7% growth year over year. This decline is also apparent in the

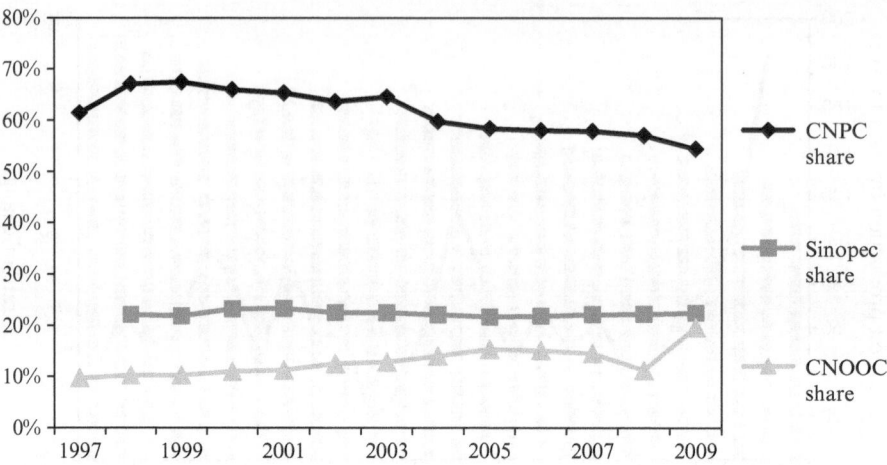

Fig. 2.3 China's top three petroleum companies' share in total crude oil production. *Source*: International Petroleum Economics (Chinese Journal)

statistics of the three largest companies. According to the CNPC annual report in 2009, CNPC crude oil production was 0.8 billion barrels, falling by 3.1% year over year and natural gas for sales 2.1 billion cubic meters, increasing by 13.3% year over year; CNPC overseas crude oil production reached 70 million tons, increasing by 12.5% year over year, and overseas natural gas 10 billion cubic meters, increasing by 50% year over year. For Sinopec, its production of crude oil and natural gas was 0.3 billion barrels and 299 billion cubic meters respectively, with increases of 1.5 and 2% separately. For CNOOC, its output of oil and gas reached 0.2 billion barrels of oil equivalent, increasing by 17.2% year over year. These three companies have played a leading role in national crude oil production, and since 1997 the yield of crude oil from CNPC has constituted for more than half of the production (Fig. 2.3).

Of all the large oilfields in China (Table 2.5), Daqing oilfield has played a leading role in production with a yield of 50 million tons annually for many years (Sun 2010). Since 2003, Daqing has entered into a new, "sustainable development" stage, producing more than 40 million tons annually and continuing to provide a reliable resource guarantee for China's energy security. In 2009, Shengli oilfield had an annual output of 27.8 million tons; its output has remained above 26.6 million tons since 1999. With its oil production exceeding 15 million tons, Changqing oilfield has become the third largest oilfield on mainland China in 2009, following Daqing and Shengli, and it is also the largest oil and gas producer in the western region of China. The production of Tarim oilfield exceeded 5 million tons for the first time in 2002, and has ranked sixth or seventh since then.

Along with China's massive economic development, China's oil consumption has increased year after year, especially, in the high energy consumption secondary industries. More energy availability has been translated into the improvement of

Table 2.5 Oil production of the top 10 oil fields in China from 1998 to 2009 (million tons/year)

Oil field name	1998	1999	000	2001	2002	2003
Daqing	55.7	54.5	53.0	51.5	50.1	48.4
Shengli	27.3	26.7	26.8	26.7	26.7	26.7
Changqing	4.0	4.3	4.6	5.2	6.1	7.0
Xinjiang	8.7	9.0	9.2	9.7	10.1	10.6
Liaohe	14.5	14.3	14.0	13.9	13.5	13.2
Jilin	4.0	3.8	3.8	4.0	4.4	4.8
Tarim	3.9	4.2	4.4	4.7	5.0	5.3
Dagang	4.3	4.1	4.0	4.0	3.9	4.2
Huabei	4.7	4.7	4.6	4.5	4.4	4.4
Zhongyuan	4.0	3.8	3.8	3.8	3.8	3.6
	2004	2005	2006	2007	2008	2009
Daqing	46.4	45.0	43.4	41.6	40.2	40.0
Shengli	26.7	27.0	27.4	27.7	27.7	27.8
Changqing	8.1	9.4	10.6	12.1	13.8	15.7
Xinjiang	11.1	11.7	11.9	12.2	12.2	10.9
Liaohe	12.8	12.4	12.0	12.1	12.0	10.0
Jilin	5.1	5.5	5.9	6.2	6.6	5.9
Tarim	5.4	6.0	6.1	6.4	6.5	5.5
Dagang	4.9	5.1	5.3	5.1	5.1	4.9
Huabei	4.3	4.4	4.4	4.5	4.4	4.3
Zhongyuan	3.4	3.2	3.1	3.1	3.0	2.9

Source: International Petroleum Economics (Chinese Journal) (in Chinese)

China's social life, increasing ownership of personal cars, and the quickening pace of construction that characterize a well-off developing society. Recently, China surpassed Japan by becoming the world's second largest oil consumer (BP 2010). In 2009, its oil consumption reached 0.4 billion tons. Consumption has grown at a rate of 6.9% annually since 2000, while the average growth rate of crude production was only 1.8% (Fig. 2.4). In the past three decades, coal has satisfied the vast majority of China's total energy consumption requirements. In 2008, oil's share in the energy mix in China was 18.7%, dropping 4.7% from its highest of 23.4% in 2002 (Fig. 2.5).

Since becoming a net oil importer in 1993, China has raised its oil imports year after year partly because of its relatively slow growth of oil supply (Fu 2010). In 1999, net imports amounted to 33.6 million tons, a figure that increased by almost six times by 2009. Annual oil imports in 2009 reached 204 million tons, increasing by 13.9% year over year and exceeding 0.2 billion tons for the first time (Fig. 2.6). Therefore, both a decrease of domestic crude output and a growth of imports made the oil refining industry in China increasingly dependent on crude oil imports. The proportion that net crude imports accounted for refinery processing rose from 50% in 2008 to 53% in 2009 (Yuan et al. 2011).

Most of China's imported crude oil comes from the Organization of Petroleum Exporting Countries (OPEC) and that proportion has been rising over the past

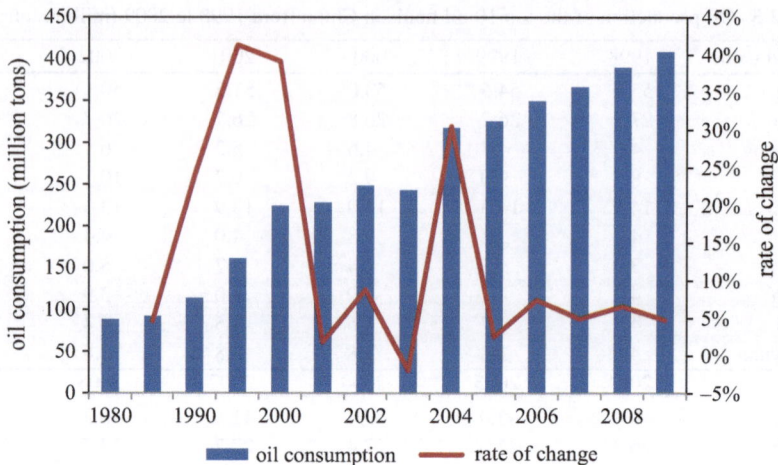

Fig. 2.4 China's oil consumption in 1980–2009. *Source*: CESY 2009, NBSC

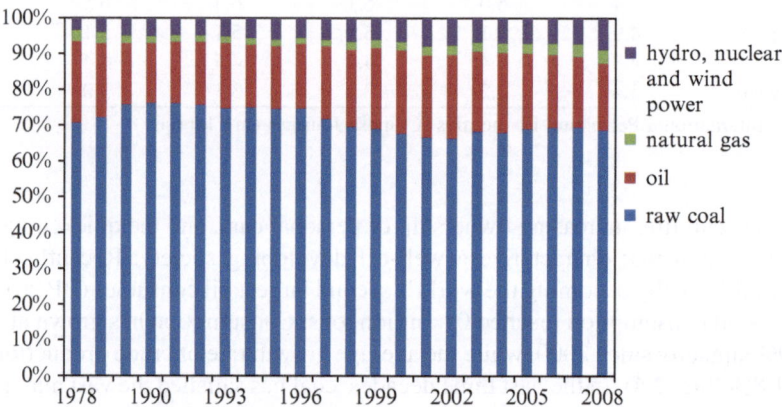

Fig. 2.5 China's structure of energy consumption in 1978–2008. *Source*: NBSC

10 years (Huang and Chen 2007). Imports, especially, from the Middle East, also have been increasing, but there has been a slight decline in the Middle East proportion. Fig. 2.7 shows the fluctuations of sources and quantities of China's crude oil from 1997 to 2009. Since 2009, 42 million tons of crude oil, accounting for 20.6% of China's total imports, has been from Saudi Arabia, making it the largest crude oil supplier for China. Angola is the second largest crude oil supplier, exporting 32.2 million tons to China, accounting for 15.8% of China's imports. Crude oil imports from Russia increased by 31.5%, but the 15.3 million tons in 2009 are still less than the imports of 16 million tons in 2006.

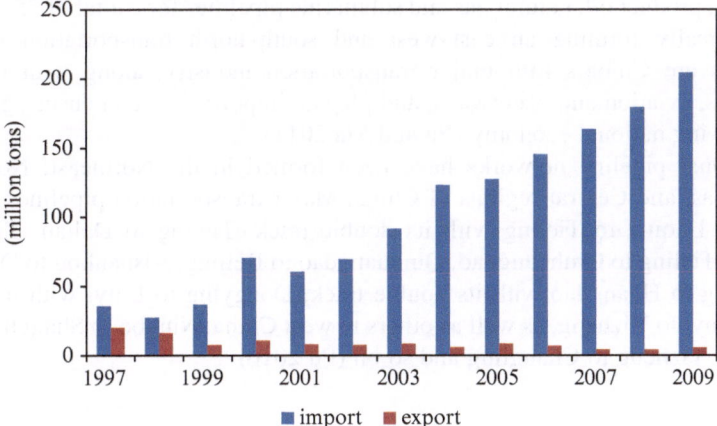

Fig. 2.6 Crude oil imports and exports of China 1997–2009. *Source*: General administration of customs

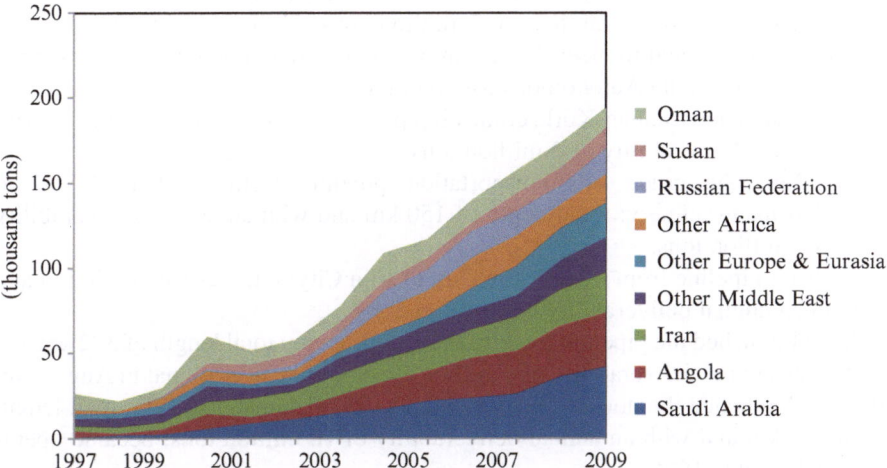

Fig. 2.7 Sources of Chinese crude oil imports 1997–2009. *Source*: General administration of customs

2.3 Development of Oil Pipelines in China

China's oil pipeline system developed along with the oil industry. Since 1988, oil and gas exploration made significant breakthroughs successively in Tuha, Tarim, Shan-Gan-Ning, Qaidam and offshore areas. Under the government policies of "stable development in the east and speeding development in the west" and "developing of oil and gas simultaneously," construction of crude oil and other kinds of pipelines began. Through 2009, CNPC, Sinopec and CNOOC have built crude oil, product oil, natural gas and submarine pipelines for a total of 75 thousand km, basically forming an east–west and south–north transportation network. Pipelines are China's fifth major transportation industry, along with railways, highways, aviation and waterways, and play an important role in ensuring a stable and growing national economy (Pu and Ma 2011).

Regional pipeline networks have been formed in the Northeast, Northwest, North, East and Central regions of China. Major transportation pipelines include: those of Daqing to Tieling with its double track, Tieling to Dalian, Tieling to Fushun, Tieling to Qinhuangdao, Qinhuangdao to Beijing, Alshankou to Dushanzi, Dongying to Huangdao with its double track, Dongying to Linyi with its double track, Linyi to Yizheng, as well as others in west China, Ningbo to Shanghai and to Nanjing, Yizheng to Changling and so on (Pu 2010).

2.3.1 China's Oil Pipeline Development (1988–2010)

1988, 6: Launched Huatugou-Gormu crude oil pipeline: total length of 435 km and with an annual deliverability of 2 million tons. In Sep. 1990, this pipeline became operational to meet the requirements of national defense and economic development in the Autonomous Region of Tibet.

1991, 7: Launched Lunnan-Korla crude oil pipeline: total length of 191 km and with an annual deliverability of 3 million tons.

1993: Built the crude oil transportation pipeline from Huabei oilfield to Shijiazhuang refinery: total length of 150 km and with an annual deliverability of 3.5 million tons.

1995: Built pipeline from Changchun City to Jilin City: total length of 156 km and with an annual deliverability of 4 million tons.

1995, 7: Launched the pipeline from Tazhong to Lunnan: total length of 302 km and with an annual deliverability of 6 million tons; became operational in Aug. 1996.

1995, 9: Launched the double track of pipeline from Lunnan to Korla: total length of 161 km and with an annual deliverability of 10 million tons; became operational in June 1996.

1996, 6: Launched the pipeline from Korla to Shanshan: total length of 476 km and with an annual deliverability of 10 million tons; became operational in June 1997.

1996: Built and put into production the pipeline from Ansai to Yan'an: total length of 107 km and with an annual deliverability of 1.22 million tons.

1997: Built the pipeline from Jingbian to Maling: total length of 200 km and with an annual deliverability of 1.1 million tons.

1997: Built the pipeline from Mabianzhou Daya Bay to Guangzhou Petrochemical Company: total length of 170 km and with an annual deliverability of 12 million tons.

2001: Built the pipeline from Ansai Jing'an oilfield to Xianyang: total length of 463 km and with an annual deliverability of 3.5 million tons.

2004, 3: Built the pipeline from Ningbo to Shanghai and then to Nanjing: total length of 645 km and with an annual deliverability of 20 million tons.

2004, 4: Launched the pipeline along Yangtze east to Yizheng Jiangsu Province and west to Changling Hunan Province: total length of 973 km and with an annual deliverability of 27 million tons; became operational in Dec. 2005.

2005: Kazakhstan-China crude Pipeline project (phase 1), China's first multinational long distance transmission pipeline was completed, west to Atasu Kazakhstan, across Alshankou China and east to Dushanzi Xinjiang Uygur Autonomous Region: total length of 1,208 km and with an annual deliverability of 20 million tons.

2005, 12: The first section (Atasu-Alshankou: total length of 962.2 km) of the Kazakhstan-China Crude Pipeline was completed and became operational.

2006: Western crude pipeline (part of western crude and product pipelines) became operational with a designed transport capacity of 2 million tons per year. This pipeline and Kazakhstan-China crude pipeline formed the Strategic West to East Oil Transmission Project. The pipeline formally became operational in June 2007.

2007: Implement the Phase II project of the Kazakhstan-China Gas Pipeline from Kenkiyak to QomCole: total length of 761 km; became operational in Oct. 2009.

2007, 6: Crude pipeline from Caofeidian off-take station to oil terminal in Tianjin became operational: total length of 190 km with a designed annual deliverability of 20 million tons.

2008, 2: Crude pipeline from Aoshandao Zhoushan City in Zhejiang Province to Cezidao in Zhoushan City became operational: total length of 45 km.

2008, 4: Started the construction of Jibai crude pipeline, north to Ji2 Union Station of oil production plant located in Jiyuan oil town of Dingbian county in Shaanxi Province, and south to Baibao oil transporting station of Wuqi county in Shaanxi Province: total length of 104 km; became operational in Dec. 2008.

2009: Construction commenced of Russia-China crude pipeline, consisting of three sections: far-east pipeline from Tayshet to Skovorodino, border pipeline from Skovorodino to Mohe and Mohe-Daqing pipeline. The pipeline is 1,030 km long. Border pipeline from Skovorodino to Mohe launched on 27 April and Mohe-Daqing pipeline on 18 May. It will be operational and Russia will begin to supply 15 million tons of crude to China annually through this pipeline for the next 20 years.

2009. 9: Launched the pipeline of Shikong-Lanzhou from the Shikong Ningxia take-off station to Lanzhou terminal station: total length of 359 km; planned to be operational in June, 2010.

2009, 9: Anbao-Yinchun pipeline started construction: total length of 141 km; planned to be operational in June, 2010.

2009, 11: Launched the pipeline from Rizhao port take-off station to Dongming Petrochemical terminal station: total length of 462 km; will be operational in June, 2011 on schedule.

In 2010, construction of key pipeline projects that include a third Shaan-Jing gas pipeline, as well as Qinhuangdao-Shenyang and Tai'an-Qingdao gas pipelines, crude pipelines such as Mohe-Dalian and Rizhao-Dongming, and product pipelines consisting of the Sunan and Pearl River Deltas (phase 2) projects. From 2010 to 2015, 50,000 km of pipelines are expected to be completed. China's total oil and gas pipelines will reach 110,000 km by 2015 (Pu and Ma 2011). Meanwhile, a large quantity of auxiliary projects include underground gas storage, liquefied natural gas (LNG) receiving stations and reserve storage will be constructed to guarantee the security of oil and gas supplies. By 2015, the nationwide network system will be completed which is characterized by diversified sources, flexible transmission, adequate logistic support and stable supply.

The next few years are likely to be the peak time for China's oil/gas pipeline construction. As the link between resources and markets, pipeline construction will make considerable development possible with large imports of foreign resources, increasing reservoirs and rising output from domestic oilfields, and vigorous development of regional markets. China will focus on gas development in the plan for the next 5 years (Qu 2011).

2.4   Refining and Marketing Oil in China

China's crude oil processing capacity increased from 1998 to 2009. In 2009, total annual crude oil refining throughput was about 375 million tons, increasing by 7.9% year over year. For CNPC, throughput was 112 million tons, falling by 2.5% year over year, and for Sinopec, 183 million tons, rising by 6.7% year over year (Fig. 2.8).

From 1998 to 2009, the production of major refined products such as gasoline, kerosene, diesel, lubricating oil, and fuel oil generally showed an upward trend. The production of gasoline, kerosene and diesel in 2009 increased by 107%, 157% and 189% respectively compared with production in 1998. In 2009, total output of refined oil reached 228 million tons, increasing by 9.4% year over year, including 72 million tons of gasoline, and increasing 13.1% year over year, 15 million tons of kerosene, increasing by 27% year over year and 141 million tons of diesel, increasing by 6% year over year (Fig. 2.9).

China has always attached importance to crude refining. A large number of projects, that were put into production or commenced and developed, have achieved remarkable results. In 2009, contrary to world trends, China's crude refining industry grew and established large-scale refining installations at Huizhou Guangdong, Fujian, Dushanzi and Tianjin, etc. (Song 2010). Primary processing capacity increased by 45 million tons and total crude processing capacity rose to 483 million tons per annum, making China the second-largest refining country behind the United States. Meanwhile, Sinopec China and CNPC became the world's third-largest and eighth-largest refining companies respectively (Table 2.6).

By the end of 2009, there were 17 refineries that reach the level of refining 10^7 tons per year in China, including Sinopec with 11, CNPC with 5 and CNOOC with 1.

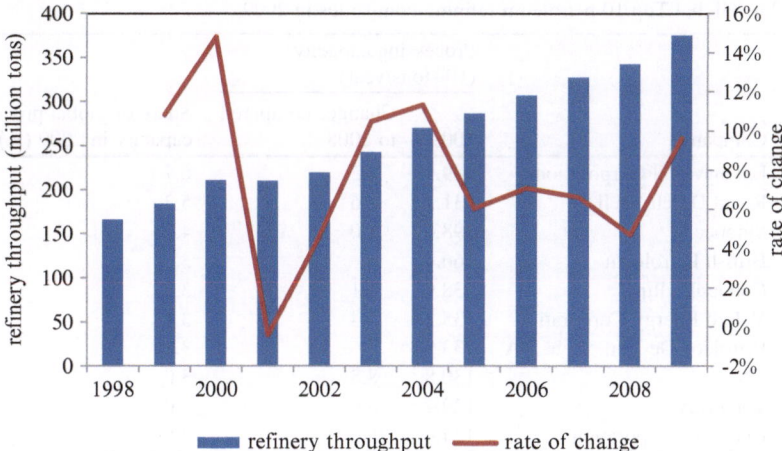

Fig. 2.8 China's crude oil refinery throughput in 1998–2008. *Source*: International Petroleum Economics (Chinese Journal)

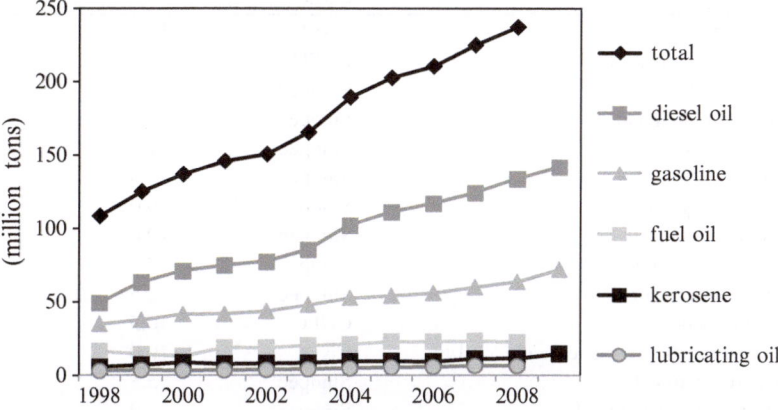

Fig. 2.9 China's major refined products production in 1998–2008. *Source*: Data Handbook for Energy Planning, International Petroleum Economics

The average scale of refining bases for Sinopec was 6.4 million tons per year, and for CNPC, 5.1 million tons per year (Qian 2010) (Table 2.7).

According to the "Medium and Long Term Development Plan of Refining Industry" published in December, 2005 by the National Development and Reform Commission (NDRC), more than 20 competitive 10 Mt per year refineries would be established by the end of 2010, accounting for 65% of total processing capacity. From 2009 to 2015, the total newly increased refining capacity is expected to come up to 197 Mt per year, including Sinopec, 107 million tons per year, CNPC, 80 Mt per year and CNOOC, 10 million tons per year.

Table 2.6 Global Top 10 petroleum refining companies in 2009

Rank	Company	Processing capacity (10^6 tons/year)		Share of global processing capacity in 2009 (%)
		2009	Changes compared to 2008	
1	ExxonMobil Corporation	289.9	8.3	6.7
2	Royal Dutch Shell	231.5	1.6	5.3
3	Sinopec	198.6	8.0	4.6
4	British Petroleum	166.4	0	3.8
5	ConocoPhillips	138.9	4.1	3.2
6	Valero Energy Corporation	135.2	5.4	3.1
7	Petroleos De Venezuela S.A	133.9	0	3.1
8	CNPC	130.8	8.8	3.0
9	Total S.A	129.7	−3.0	3.0
10	Chevron Corporation	124.6	25.6	2.9

Source: International Petroleum Economics (Chinese Journal) (in Chinese)

Table 2.7 Refining bases with the level of 10^7 tons per year in China (million tons/year)

Base	Owner	2005	2010
Dalian Petrochemical	CNPC	10.5	20.5
Fushun Petrochemical	CNPC	10.0	10.0
Yanshan Petrochemical	Sinopec	8.0	10.0
Shanghai Petrochemical	Sinopec	14.0	14.0
Gaoqiao Petrochemical	Sinopec	11.0	11.3
Jinling Petrochemical	Sinopec	13.0	13.5
Zhenhai Petrochemical	Sinopec	20.0	20.0
Qilun Petrochemical	Sinopec	10.0	10.0
Guangzhou Petrochemical	Sinopec	7.7	13.0
Maoming Petrochemical	Sinopec	13.5	13.5
Lanzhou Petrochemical	CNPC	10.5	10.5
Dalian West Pacific Petrochemical Co., Ltd.	CNPC	10.0	10.0
Tianjin Petrochemical	Sinopec	5.5	15.0
Fujian Refining & Chemicals	Sinopec	4.0	12.0
Dushanzi Petrochemical	CNPC	5.5	10.0
Qingdao Refinery	Sinopec	–	10.0
Huizhou Refinery	CNOOC	–	12.0

Source: International Petroleum Economics (Chinese Journal) (in Chinese)

2.5   Refined Oil Market

2.5.1   Supply and Demand of Refined Oil

Since 2003, China's consumption of petroleum-derived products has become almost equal to its refining capacities under the two-way adjustment plan (import and export). Gasoline exports accounted for only a small proportion of gasoline

Table 2.8 China's supply and demand of petroleum-derived products 2003–2008

Petroleum products (million tons)	Year	2003	2004	2005	2006	2007	2008
Gasoline	Supply	40.7	47.0	48.6	55.9	56.0	63.5
	Consumption	40.7	47.0	48.5	52.4	55.6	63.4
Diesel	Supply	84.7	99.5	109.7	116.5	123.4	133.2
	Consumption	84.1	99.0	109.7	118.4	124.3	138.9
Kerosene	Supply	9.2	10.6	10.7	9.7	11.6	11.7
	Consumption	9.2	10.6	10.8	11.3	12.4	12.8

Source: China Energy Statistic Yearbook, China Statistic Yearbook (in Chinese)

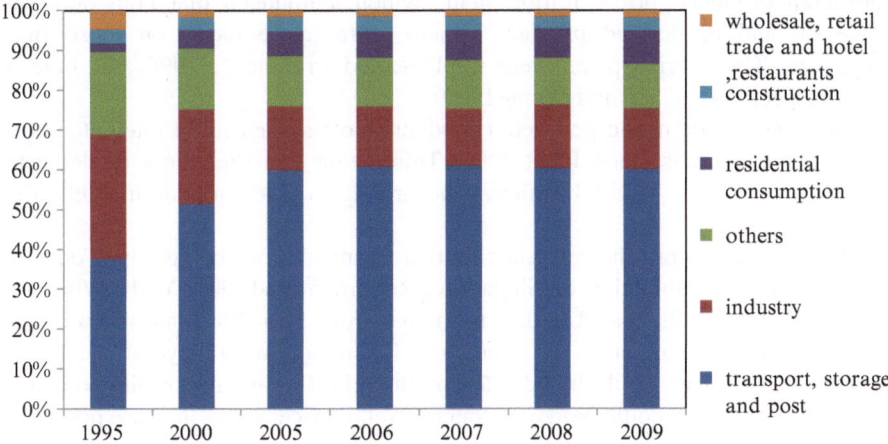

Fig. 2.10 Petroleum products consumption by sector in China. *Source*: CESY 2010

production. Demand for diesel and kerosene relied mainly on domestic output. Demand for petroleum products grew rapidly from 2003 to 2008. Average annual growth rate of gasoline, diesel and kerosene demands were 9.3%, 10.6% and 6.8% respectively (Table 2.8).

China's petroleum-derived products are consumed by all components of China's economy: agriculture, forestry, animal husbandry, fishery, water conservancy, industry, construction, transportation, warehousing, postal service, catering and household consumption, etc. (Fig. 2.10).

The growing demand for oil-derived products in transportation, warehousing and postal service motivated growth of gasoline and diesel production in China. Because of the rapid development of transportation, total turnover of freight traffic was 12 trillion ton per km in 2009, up 127% compared to 2003. The volume of passenger transportation was 2 trillion persons per km in 2009, an increase of 79.8% compared to 2003. Improvement of roads and other infrastructure construction, and growth of population and vehicle sales promoted a sharp rise in diesel consumption in China.

2.5.2   Price Mechanism of Crude Oil and Refined Oil

China's crude oil price has been integrated with international crude oil markets, while the same case does not apply to oil-derived products. Since the adjustments to the price of oil products effects a wide range of industries and people's livelihoods, the price has been regulated by the government. Since 1998, China gradually accelerated the pace of price adjustment (usually increases) on crude oil and oil-derived products.

China's oil price started to integrate with the international market since 1998. On June 3, 1998, the State Planning Commission promulgated a "crude oil and petroleum products' prices reform plan" which announced that both domestic crude oil and oil-derived products' prices were to be based on those from Singapore. These price plans were implemented in June 1, 1996 and June 5, 1996, respectively (Chen and Wang 1999).

The prices of domestic petroleum products have started to be integrated with international markets since June, 2000. This meant that the prices of domestic petroleum products would be altered according to international market price changes.

From that point on, the reference market changed from being based only on Singapore to a combination of Singapore, Rotterdam and New York. When the international price fluctuated within the range from 5 to 8%, domestic oil price would not be altered, however, when the fluctuation went beyond the range, medium retail prices would be adjusted by the NDRC. Two major oil companies (Sinopec and CNPC) could work out concrete retail prices using the base price plus or minus 8%.

In January of 2007, a new refined oil pricing mechanism "crude oil plus cost," was implemented. The system, first, took an average of Brent, Dubai and Minas crude oil prices as a benchmark. Second, refining costs, reasonable margin, domestic tariffs and circulation fees for refined oil, etc., were added from the domestic oil retail base price. Unfortunately, this mechanism led to serious lags in price alterations.

A newly reformed price mechanism began in Jan. 1, 2009. Domestic onshore oil prices continued to be integrated directly with international markets; domestic refined oil products continued to be priced in coherence with international markets indirectly and with governmental controls. Domestic oil products (gasoline and diesel) would be priced on the basis of international crude price, with domestic average processing cost, tax revenue and reasonable refining margins taken into account. Finally, the new system changed gasoline and diesel's retail price from "base price permitted" fluctuations into "maximum retail price (Feng et al. 2009)."

There were 20 price alterations from Jan. 2003 to June 2009. China raised fuel prices 14 times and lowered prices 6 times. Refinery gate prices for gasoline and diesel increased 5,000 Yuan/ton and 4,440 Yuan/ton, respectively (Table 2.9).

Table 2.9 Wellhead price and alterations, China, 2003–2009

Times	Date (mm/dd/yy)	Refinery gate price (Yuan/ton)	Change %
1	1-1-2003	3,020	–
2	2-1-2003	3,210	6.3
3	5-10-2003	2,920	−9.0
4	12-6-2003	3,210	9.9
5	3-31-2004	3,510	9.4
6	8-25-2004	3,750	6.8
7	3-23-2005	4,050	8.0
8	5-24-2005	3,900	−3.7
9	6-25-2005	4,100	5.1
10	7-23-2005	4,400	7.3
11	3-26-2006	4,700	6.8
12	5-24-2006	5,200	10.6
13	1-14-2007	4,980	−4.2
14	11-1-2007	5,480	10.0
15	6-20-2008	6,480	18.3
16	12-19-2008	5,580	−13.9
17	1-15-2009	5,440	−2.5
18	3-25-2009	5,730	5.3
19	6-1-2009	6,130	6.9
20	6-30-2009	6,730	9.8

Source: National Development and Reform Commission, http://www.sdpc. gov.cn/

2.6 International Cooperation

2.6.1 Development of International Cooperation of Chinese Petroleum Companies

After more than 10 years of development, China's oil companies have made great progress in international business. Most projects focused on comprehensive efforts, including oilfield production, technical services, refinery and pipeline construction, etc. Increasingly business transformed from the extraction processing business into capital management. Four strategic development zones were established: the Middle East-North Africa, Central Asia-Russia, South America, and South Asia. By 2009, 69.6 million metric tons of crude oil and 8.2 billion cubic meters of natural gas had been produced by CNPC through 81 oil and gas projects located in 29 countries. This constituted a historic breakthrough in oil and gas production for China. The scale of overseas operations has been enlarged continuously. Production capacity of crude oil and natural gas by the company CNPC reached 70 million tons per year and 10 billion cubic meters per year.

China's oil and gas production that was produced in other countries exceeded 0.1 billion tons for the first time in 2008. Meanwhile, China's share of this production exceeded 50 million tons for the first time. Based on the achievements in 2008,

Table 2.10 CNPC's main business development distribution in the world

Location	Countries
Asia	Iran, Oman, Syria, Pakistan, Indonesia, Burma, Kazakhstan, Turkmenistan, Uzbekistan , Mongolia, Thailand
Africa	Sudan, Algeria, Mauritania, Tunisia, Libya, Chad, Niger
Europe	Russia, Azerbaijan
America	Canada, Mexico, Venezuela, Peru, Ecuador

Source: Annual Report of CNPC

China's enterprises were able to maintain stable growth in overseas operations in 2009 despite the global economic problems; production reached more than 0.1 billion tons and China's share amounted to about 55 million tons. At the same time, China's business enterprises, through mergers and acquisitions, became one of the major buyers in international oil and gas markets. In 2009, 13 mergers and acquisitions were announced by Chinese companies, of which 11 turned out to be successful. The total amount of these transactions reached nearly US$16 billion, making 2009 the most active year for Chinese companies' overseas mergers and acquisitions.

From the above we can see that Chinese petroleum enterprises represented by CNPC, CNOOC, Sinopec and Sinochem Group (Sinochem), have made remarkable achievements in multinational operations, and have demonstrated that their internationalized management is reaching a mature stage.

2.6.1.1 Development of CNPC International Business

CNPC is the first large state-owned enterprise to carry out international cooperation in China. In 1993, the CNPC won their bid for production of block 6/7 in Peru's Talara region. It was not only the first overseas oilfield development project CNPC operated, but also the breakthrough for China's petroleum enterprise in international business (Qin 2007). China began to understand that it had the technical and business experience to be able to play on the international stage.

Presently, CNPC owns and operates more than 70 overseas projects, basically distributing in more than 30 countries, such as Iran, Sudan (Table 2.10). Projects in the Middle East and North Africa consist of projects in Block 1/2/4, Block 6 and Block 3/7 in Sudan, a project in Oman, a project in Algeria, and a project in Syria. Projects in Central Asia-Russia include: AktobeMunai Gas, Kenkiyak-Atyrau Pipeline, North Buzachi, and Central Block at the Eastern Edge of the Precaspian Basin in Kazakhstan, a project in Pakistan, a project in Gumdak in Turkmenistan, projects in Gobustan and in Azerbaijan. Projects in South America include: Block 6/7 in Peru, a project in Block 11 in Ecuador, and projects in Intercampo, Caracoles and Orimulsion in Venezuela.

At present, international cooperation is becoming a new economic growth point for CNPC. Table 2.11 highlights the rapid development process of CNPC in international operations (CNPC 2009):

Table 2.11 Major Events that have happened during the development of CNPC

Year	CNPC events
1993	(a) Obtained exploration and service assignments on Block 7 in Peru's Talara oilfield (b) Produced the first barrel of crude oil overseas through the project in Canada
1994	(a) The exploration contract with Papua New Guinea was signed
1995	(b) CNPC won the tender of exploration project of Block 6 in Muglad Basin and signed production sharing agreement
1997	(a) CNPC signed production sharing agreement for Block 1/2/4 in Muglad Basin Sudan and thereafter achieved partial equity of Block 3/7 (b) CNPC signed the contract in which Khartoum Refinery in Sudan was jointly invested in and constructed by CNPC (c) CNPC acquired a 60 % stake in AktobeMunai Gas by paying out US$0.32 billion and achieved mining rights in Aktobe oilfield (d) CNPC signed a production sharing agreement for Al-Ahdab oilfield (e) CNPC won tenders for the Intercampo and Caracoles oilfields in Venezuela (f) Obtained exploiting rights for Uzen oilfield in Kazakhstan
1999	(a) The first shipment of oil from Block 1/2/4 Sudan was exported
2000	(a) Khartoum refinery with a processing capacity of 2.5 million tons per year became operational
2001	(a) The first overseas service station was built in Khartoum (b) Signed contracts for projects in Orimulsion, Venezuela and in Burma
2002	(a) Project of polypropylene in Sudan produced qualified products (b) CNPC acquired Devon Energy's share in Indonesia for US$0.2 billion (c) Acquired Petroleum Development Oman for US$25 million (d) Signed development contract with Turkmenistan, contract for project of K&K in Azerbaijan and contract for Central Block at the Eastern Edge of the Precaspian Basin in Kazakhstan (e) Kenkiyak-Atyrau Pipeline became operational
2003	(a) Investment in a project in Block 1/2/4 in Sudan was paid back (b) Signed contracts for Gobustan oilfield in Azerbaijan, the Adrar Upstream and Downstream Integrated Project, the North Buzachi oilfield project with Kazakhstan and projects in Syria and Ecuador successively (c) Won a project in the Gobustan oilfield in Azerbaijan (d) CNPC signed a development and production enhancement contract for the Kabiba oilfield with Syrian Petroleum Corporation (SPC) (e) CNPC acquired 50 % equity in Amerada Hess Indonesia Holdings Ltd. (f) Signed a contract with the Algerian state-owned Hydrocarbons Company, and obtained the upstream and downstream integration contract for oilfield development and construction, and operating the refinery in Adrar province (g) CNPC purchased 35 % stake of Saudi Arabia Companies-North Buzachi oilfield in Kazakhstan (h) Acquired Block 11 in Ecuador (i) CNPC purchased a 65 % share of North Buzachi in Kazakhstan that was owned by Chevron Texaco of United States
2004	(a) CNPC inked an agreement on stock rights assignment for the Iranian MIS project with Canadian Sheer Energy Company giving CNPC 49 % of the project (b) Won the tender for Block 438b in Algeria (c) CNPC signed a contract with the Mauritanian Ministry of Industry and Mining for exploration and development of Block Ta13, Block Ta21 and Block 12 (d) CNPC acquired the NK exploration block and a 50 % holding in the SLK oilfield from Kuwait Foreign Petroleum Exploration Company

(continued)

Table 2.11 (continued)

Year	CNPC events
	(e) CNPC signed an agreement with Ay-Dan to buy shares of ADM and acquired a 100 % stake in it. CNPC also acquired 50 % shares in Konys and Bektas oilfields
2005	(a) CNPC acquired PetroKazakhstan with a total investment of US$4.18 billion. It was a typical multinational acquisition and the largest overseas merger for a Chinese petroleum enterprise, which led to an increase in crude production of 10 million tons per year
2007	(a) CNPC and Venezuela signed an agreement for joint exploration of the Zumano oilfield by Petrozumano. It was the most attractive achievement since 2004, when China started oil exploration in Latin America
	(b) CNPC signed a production sharing contract and a natural gas sales-and-purchase agreement with Turkmenistan. According to the agreements, Turkmenistan will export 30 billion cubic meters of natural gas to China annually for 30 years
2008	(a) A cooperation project was launched with Chevron for the development of acid gas in northeastern Sichuan Province, which was the biggest continental upstream cooperation project of China
2009	(a) Acquired a 45.5 % stake in SPC (one of the top three refineries of Singapore). SPC became a new platform for CNPC's international business
	(b) CNPC and BP jointly won a service contract to develop Iraq's largest Rumaila oilfield
	(c) CNPC, Total and Petronas won the Halfya oil field (in Iraq) service contract; signed a series of agreements on oil and gas cooperation with Turkmenistan, Kazakhstan, Uzbekistan and Russia; signed a cooperation contract with Canada's Athabasca Oil Sands Corporation for oil sands exploration and development; purchased Mangistau Munai Gas for US$3.3 billion, further expanding the cooperation business in Central Asia; finally, CNPC and the Costa Rica National Oil Company set up a joint venture company to upgrade and expand Costa Rica's MOIN refinery

Source: Annual Report of CNPC

2.6.1.2 Development of Sinopec International Business

Sinopec has paid great attention to international operations. The Group accelerated its "Going Global" pace and achieved great progress. Presently, the Group has cooperation agreements with many countries, including Iran, Saudi Arabia, Gabon, Kazakhstan, Yemen, Ecuador, Angola, Oman, Libya, Niger and Mali, to exploit and develop their local oil and gas projects (Sinopec 2009).

Major events taking place in the internationalized management process of Sinopec and its controlled companies are given in Table 2.12.

2.6.1.3 Development of CNOOC's International Business

On January 30, 1982, the Chinese State Council promulgated "Regulations of the People's Republic of China Concerning the Exploitation of Offshore Petroleum Resources in Cooperation with Overseas Partners," legislatively granting the CNOOC an exclusive right of conducting oil exploration, development, production and sales offshore from China, and giving CNOOC complete charge of the business

Table 2.12 Major events that have happened during the development of Sinopec

Year	Sinopec events
2000	(a) Signed exploration and service contract for Kashan Block with the Ministry of Petroleum of Iran
2001	(a) Signed an agreement for Yemen S2 exploration and development project with German Prussian Shige Company of Germany
2002	(a) Signed a cooperation contract for the exploration of Zhaerzhating oilfield in the Eastern Sahara desert with the Algeria state-owned Hydrocarbons Company
2003	(a) Successfully drilled a high-yield well in risk exploration of Iran's Kashan Block
	(b) Purchased Ecuador Block 16 and thus owned the first non-operating project in South America
2004	(a) Sinopec, together with the Saudi Arabia Oil Company, signed a Natural Gas Exploration and Development Agreement upon Section B in Rub Alkhali Basin with the Ministry of Petroleum of the Kingdom of Saudi Arabia. It had been 10 years since China implemented the strategy of "Going Global" and this was China's first official project on the world's largest provider of oil and gas resources, Saudi Arabia
2006	(a) Purchased shares of Angola's three deep-sea oilfields with a total investment of US$0.69 billion, including: a 27.5 % share of Block 17, a 40 % share of Block 18 and a 20 % share of Block 15
	(b) Successfully purchased Udmurtneft Petroleum Corporation of Russia with an investment of US$3.5 billion
	(c) Won six bids for international refining engineering and service projects with total contract value of US$3.1 billion; obtained an upgrading project for Iran's Allah refinery with the total contract value of 2.2 billion Euros, which was by far China's biggest overseas refining engineering and service project
2009	(a) Purchased a 10 % stake in the "Light of North" oil sands project in Alberta Province Canada from Total. This gave Sinopec and total half of the shares in the project. The projected recoverable oil reserves from this project amount to 1 billion tons
	(b) Purchased Addax Petroleum Corporation for US$8.3 billion—by far the largest overseas acquisition for a Chinese enterprise. Addax is an independent oil company, with 25 exploration blocks. Its oil and gas resources are concentrated in Nigeria, Gabon and Iraq's Kurdish region. By the end of 2008, its 2P reserves reached 0.5 billion barrels; its average daily output in 2008 was about 140 thousand barrels

Source: Annual Report of Sinopec

of exploiting oil resources in cooperation with overseas partners. On February 15, 1982, the CNOOC was formally established in Beijing. The important task of cooperating with overseas partners and developing Chinese offshore oilfields, CNOOC has played an important role in the process of "Going Global" for Chinese enterprises (Luo 2004). From its very inception, CNOOC has deepened its international operations, achieved great development abroad, and expanded its skills in overseas business.

In recent years, CNOOC has continuously expanded its international business. Its internationalization has been accelerated through mergers and acquisitions, which made contributions to the rapid growth of its overseas output. Major events that have happened during this process are given in Table 2.13.

Table 2.13 Major Events that have happened during the development of CNOOC

Year	CNOOC events
1994	(a) CNOOC purchased 32.6 % of ARCO's (USA) rights in Malacca Strait block in Indonesia with an investment of US$16 million
1995	(a) Purchased a 6.9 % stake in the Malacca Strait block in Indonesia from Nippon Oil Corporation
2002	(a) Purchased three blocks of oil and gas fields from Australia and Indonesia with an investment of US $1.2 billion
	(b) With an investment of US$0.59 billion, CNOOC purchased partial rights in Indonesia's five oilfields from Spain's Reposol, and became the largest offshore petroleum company in Indonesia
	(c) Built a sharing partnership with Indonesia in Tangguh oilfield's upstream products, and signed a 25-year liquefied natural gas (LNG) sales and purchase agreement. The contract value amounted to $8.5 billion
2004	(a) Acquired Australia's northwest shelf natural gas project, and signed a 25-year LNG supply contract with this project's operators
2005	(a) Ventured into oil sands resource development for the first time and purchased an 18.5 % share of MEG company in Canada
	(b) On behalf of China, CNOOC has conducted international bidding for cooperative exploitation of oil and gas; 172 petroleum agreements and contracts have been signed with 75 companies belonged to 21 countries and regions; Currently, CNOOC's overseas assets are distributed in the Asia-Pacific, Africa and North America regions. Professional service companies affiliated with CNOOC also have ties with foreign markets involving several regions, such as Southeast Asia, America, Africa and the Middle East
2006	(a) CNOOC acquired a 45 % equity interest in OML 130 Offshore Nigeria for USD 2.3 billion
	(b) CNOOC's 27 overseas projects have been in operation and the company's total investment reached USD 4.8 billion; overseas oil and gas net proven reserves are 0.3 billion barrels of equivalent. Overseas output amounted to 5.8 million tons of oil equivalent, accounting for 14.4 % of CNOOC's total domestic and foreign output. Risk exploration prompted CNOOC to expand rapidly in overseas business
2007	(a) CNOOC had equity in 39 sections distributed in 8 countries. Its operation covered an area of 0.2 million square km. Its risk exploration covered an area of 0.4 million square km
2009	(a) CNOOC's Akpo deepwater oilfield in OML 130 project in Nigeria successfully commenced production. CNOOC had a 45 % equity share of the OML 130 project. Akpo oilfield is one of the world's major deep-water discoveries. Total expected controlled reserves and proven reserves of this great oilfield can amount to about 0.62 billion barrels
	(b) Acquired 20, 10, 10 and 10 % equity in Tucker, Logan, Cobra and Krakatoa blocks respectively of Norway National Petroleum Corporation in Gulf of Mexico Block; The completion of the transaction meant that the investments of Chinese petroleum enterprise had entered the Gulf of Mexico

Source: Annual Report of CNOOC

2.6.1.4 Development of International Operations of Sinochem

Sinochem was founded in 1950. It was a key state-owned enterprise under the supervision of State-owned Assets Supervision and Administration Commission of the State Council. Its predecessor was China Import Co., Ltd. Its business segments include: petroleum, chemical fertilizer, chemical products trade, distribution and

Table 2.14 Major Events that have happened during the development of Sinochem

Year	Sinochem events
2003	(a) Successfully acquired Atlantis Holding Norway AS (ATLANTIS) for US$0.1 billion; It was originally a fully-owned subsidiary of Norway Petroleum Geo-Service ASA (PGS) which mainly dealt with exploration and production. Its assets include 11 contract blocks distributed in Tunisia and the United Arab Emirates, with a project operation area of 10,000 km^2
	(b) CRS Resources LDC was purchased by Sinochem Corporation from ConocoPhillips Corporation; It is located in Ecuador. Its assets include 14 % of the rights in Ecuador Orient Basin Block 16, whose proven recoverable reserves amount to 4.6 million ton
2006	(a) Held high-level talks with the Indian state oil corporation. Both sides planned to cooperate on crude oil, oil products trade, upstream exploration and development, and chemical marketing, to realize win-win cooperation for both countries
2007	(a) Acquired the New XCL-China, LLC ("New XCL"). The headline purchase price was approximately US$218 million and the total consideration paid was approximately US $228 million; New XCL was a privately owned Delaware Limited Liability Company whose sole asset was a non-operated 24.5 % stake in Zhao Dong petroleum block in Bohai Bay, offshore China
2009	(a) Sinochem acquired Emerald Company. Emerald owned equity ranging from 50 to 100 % respectively in 11 blocks in Syria, Colombia and Peru. In ten of these blocks, Emerald had a recoverable reserve of 100 million barrels plus 1 billion barrel of potential reserve for exploration. This acquisition signaled that Sinochem had rolled out a strategic layout of oil and gas assets in South America and the Middle East

Source: Annual Report of Sinochem

logistics, crude oil, fuel oil, natural rubber futures; overseas oil and gas exploration and development, oil refining, chemical ore mining and refining, chemical fertilizer and chemical production; also hotel and real estate development and management (Sinochem 2009). It is one of the four largest oil companies in China. Sinochem is China's largest fertilizer importer and phosphate and compound fertilizer producer, and China's most important chemical service provider.

To meet the national energy supply needs of China, Sinochem has expanded its petroleum business into foreign markets since 2002. Major events taking place in the Sinochem's internationalized management process as given in Table 2.14.

While the scope and extent of these operations may come as a surprise to many, it is no different from how many large oil consuming nations spread their risk and prepare for the future. In general China holds a minority stake in all the operations it takes part in, and like other large nations attempt to insure its future resource flows through what is normally considered legitimate international business approaches. Partner companies usually get about half of the oil developed. Nevertheless some important problems remain, including the determination of boundaries between maritime countries.

2.7 Role of Technology in the Petroleum Industry of China

In the past 20 years, China's petroleum industry has made numerous important scientific and technological achievements which are the precious wealth and intangible assets for the development of science and technology. They represent the technological level, technological strength and technological power of China's petroleum industry, and have both promoted the development and changed the outlook of that industry (Ming and Yunm 2006).

2.7.1 Achievements of Research on Geological Theory in China

From 1986 to 2009, two important contributions were made in the theory of petroleum generation: (1) the concept of immature oil was introduced which changed research on hydrocarbon generation mechanisms and organic geochemistry; (2) a systematic summary was made about the character of Jurassic coal in Tuha basin, and a multi-stage hydrocarbon-generating and oil formation model in coal measure strata was built providing the basis for oil-finding in the extensively distributed coal measure strata of China.

2.7.2 Achievements of Exploration Technologies in China

China is a complex area geologically, with many mountains and folded rocks. This has led to China making many contributions to understanding such regions. For example, the following are some major research activities of the two major companies.

1. CNPC: Explored lithological reservoirs and developed sequence stratigraphy and efficient techniques for trap identification, promoting the understanding of the regulation of lithological reservoir formation; improved exploration techniques for foreland basins, enriched and developed the theory of fault-related folds and structure modeling techniques, deepened the understanding of structure character of foreland basin and oil and gas reservoir-forming rule, and matched acquisition technology of mountain areas and fine structural imaging techniques; developing the exploration technique for carbonate rock of marine facies and volcanics, developed techniques of identification, evaluation and prediction of paleokarst reservoirs, oolitic beach reservoirs and volcanic reservoirs. Finally they deepened the understanding of hydrocarbon accumulation rule; techniques of seismic fine imaging, underbalanced drilling and complex lithological logging evaluation and reservoir reconstruction.
2. Sinopec: In the eastern fault basin, their geologists gradually formulated reservoir-forming theory of the lithological and stratigraphic oil and gas reservoirs, and instructed the oil and gas exploration and development of many basins in eastern Central Europe; built several kinds of buried-hill reservoir-forming models for weathering crust buried hill, fault block buried hill and buried hill episode; a complete set of exploration techniques was formulated for improving the probability of finding oil and gas resources, ranging

from seismic imaging to fine interpretation, and prediction and evaluation of carbonate reservoir, and various projects from discrimination of oil and gas to acid-fracturing transformation and tests in reservoir beds, underbalanced drilling techniques, horizontal drilling techniques and diameter measuring and drilling techniques; deepening the understanding of multiplex control models of hydrocarbon accumulation and growth mechanisms of high-quality reservoirs, changed the exploration fields from structural reservoir s to structural lithological reservoirs.

2.7.3   Achievements of Other Technologies in China

1. Oil and gas development: The design and application of polymer molecules, which, when first applied, resulted in substantial improvement of oil recovery factors. After 6 years researching, inventing and applying two kinds of cheap, efficient and pollution-free polymers with heat resistance and salt tolerance were developed, including comb-shaped polymers and hydrophobic associating polymers, in oil and gas field development. The remarkable achievements also were made in enhancing recovery factor of high water cut oilfields and a breakthrough in the economic development limit of steam drive for medium-deep heavy oil reservoir and ultra-low permeability reservoirs, strengthening overpressure, low permeability, sulfur and loose reservoirs, and improving exploitation in condensate gas reservoirs. These discoveries made a set of difficult-to-produce reserves of China petroleum industry gradually turn into economic recoverable reserves.

2. Drilling engineering technologies: Advances and achievements of drill engineering technologies included a new type of domestic Measurement While Drilling (MWD) and positive pulse generator of key components earth guidance system and steering systems for rotary drills; developing the technology for near drill bit resistivity and natural gamma MWD; the successful application of drilling technology for multi-lateral wells in China.

3. Geophysical exploring technologies: the improvements include software development, technological integrative matching and equipment development for geophysical exploration and well logging, and also the development of a new type of large-tonnage vibroseis-KZ-28, portable mountainous area drilling rig/drilling tool and auger mining for deep wells. These have satisfied the requirements of domestic and foreign exploration of complex areas, improved the quality of original data and information, increased production efficiency and reduced the exploration cost.

4. Oil and gas pipelines: the four core technologies for oil-gas pipelines generates, includes comprehensive engineering technology for "West to East Gas Transmission"; optimization technology of welding parameters of X80 pipeline steel and selection technology for welding material with high toughness; internal inspection technology for pipelines; and pipeline operation

technology. These technologies played an important role in accelerating the oil and gas exploitation process (Shi 1995).

5. Ground engineering technologies: China undertook research on unheated technology for paraffin base crudes in alpine region, formulating six sets of oil-field ground process models and two sets of gas-field ground process models; developed and applied five sets of low temperature separation equipment to domestic and foreign oilfields; developed technology for de-acidification of heavy oil, and provided powerful technical support for overseas oil and gas development. For the aspect of oilfield maintenance and operation engineering and technology, imported equipment were mastered, and matching technology research was strengthened and ten technology series were established, consisting of overhauling, side tracking and fracture acidizing, etc.

2.8 Energy Conservation, CO_2 Emissions Reduction and Renewable Energy

From the quantitative point of view, the quantity of carbon emissions of a country depends on the total amount and structure of energy consumption (Sun and Zhou 2011). From an economic point of view, it depends on the total size of the national economy and the efficiency of energy distribution and utilization. The efficiency of energy consumption has significant differences under different systems. Take the planned economic system as an example: in the past with the continuing development of the economy, energy consumption and carbon emission rose in tandem, and overall emissions were on a high level. However, with the deepening of market reforms, the implementation of energy conservation measures, and the adjustment of the production structure, China's energy efficiency has been improved greatly, and carbon emissions have shown a downward trend since 1978. Therefore, choosing an effective economic system (or perhaps choosing to regulate it better) to improve the efficiency of energy consumption should be an important and feasible way of carbon emissions reduction (Fig. 2.11).

According to the Energy Research Institute of NDRC, 2,146 kg of carbon dioxide is discharged when one ton of oil is burnt (Table 2.15) (Tian 2009). Thus we can calculate China's annual total amount of carbon dioxide discharged by oil combustion from 1953 to 2007. Since the rise of total energy consumption and the increase in the proportion of oil consumed, the annual total amount of carbon dioxide discharged by oil combustion has been increasing. In 1953, oil consumption was 3.8% of total energy consumption and all carbon emissions were 441 million tons. In 1978, oil consumption reached 22.7% of all energy, carbon emission reached 278 million tons, and then in 2007, the proportion of oil consumption reached 20%, but the total energy amount increased substantially, so the carbon emission of oil combustion reached 1140 million tonnes (Fig. 2.12).

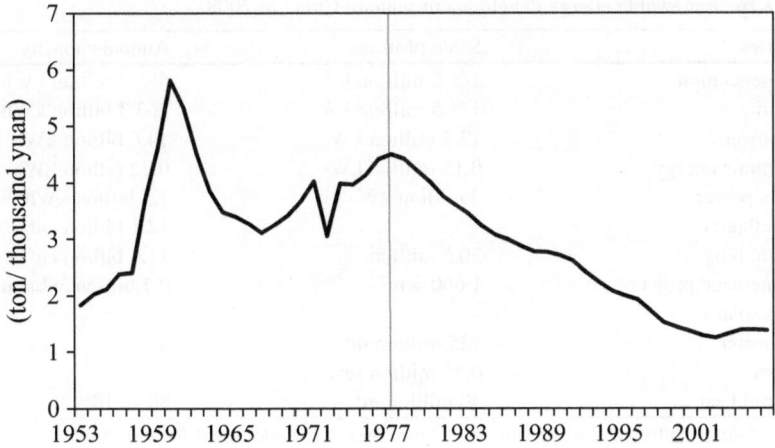

Fig. 2.11 Carbon dioxide emission per unit of economic activity in China from 1953 to 2007.
Source: CESY 1989, China Statistical Yearbook (CSY) 2006, Statistical Bulletin 2007

Table 2.15 Carbon emission factors of various energy sources (ton/ton coal equivalent (tce))

Programs	Coal and carbon	Oil	Natural gas	Hydropower, nuclear power
Fi(C)	0.748	0.585	0.444	0.0
Fi(CO₂)	2.745	2.146	1.629	0.0

Note: Fi(C) is carbon emission factor, Fi (CO₂)is carbon dioxide emission factor. The latter is obtained by the former multiplied by 3.67
Source of data: Energy Research Institute of NDRC

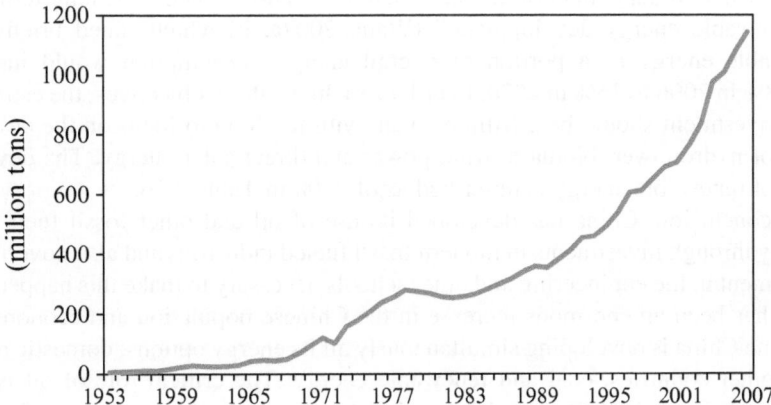

Fig. 2.12 Carbon dioxide emissions from oil consumption in China. *Source*: CESY 2010

Table 2.16 Renewable energy development scale in China in 2008

Categories	Scale of usage	Annual capacity
Power generation	186.8 million kW	585.8 billion kWh
Hydropower	171.5 million kW	563.3 billion kWh
Wind power	12.2 million kW	14.8 billion kWh
Photovoltaic energy	0.15 million kW	0.22 billion kWh
Biomass power	3 million kW	7.5 billion kWh
Gas (methane)	–	12.1 billion cubic meters
Domestic biogas	30.5 million	11.4 billion cubic meters
Large methane project	1,600 sets	0.7 billion cubic meters
Heating solar	–	–
Water heater	125 million m^2	–
Hot oven	0.45 million sets	–
Terrestrial heat	40 million m^2	80×10^{15} J

Source of data: National Energy Bureau, China Energy Development Report 2009

In recent years, the global energy supply has become tighter and tighter and the pressure on the environment has sharply increased. Soon, energy becomes a major bottleneck to China's economic development while the international pressure caused by the environment will only grow. So it is an inevitable choice that China should look for alternative sources of energy such as wind, biomass, and solar power, and develop both new, ideally renewable energy sources. Since the "Renewable Energy Law" was enacted, China has established a policy framework to support renewable energy and issued a series of supporting policies and rules (Niu 2008). In January of 2006, the NDRC promulgated such policies as: "The Guidance Catalogue of Renewable Energy Industry Development," "The regulations of renewable energy," and "The management provision of the renewable energy prices and the cost-sharing." In June, 2006, the Chinese Ministry of Finance issued "The Special Measures for Renewable Energy Development Interim Fund," while in September of the same year, the NDRC issued "The long-term plan of renewable energy development," (Wang 2007a, b) which stated briefly that renewable energy as a portion of overall energy consumption would increase from 8% in 2006 to 15% in 2020. In order to achieve these objectives, the estimated total investment should be 2 trillion Yuan, with the fund to focus on the development of hydropower, biomass, wind power and direct solar energy. The development of renewable energy is described as of 2008 in Table 2.16.

In conclusion, China has developed its use of oil and other fossil fuels enormously through investments in modern fossil fueled industries and also government investment in the engineering and other schools necessary to make this happen. The result has been an enormous increase in the Chinese population and economy. At this time China is developing simultaneously all its energy options: domestic oil, oil from other countries, coal and renewable energy. The critical role of oil is well known to many authorities and is receiving precedence despite large efforts the production of domestic oil may be reaching a plateau. Consequently China has established oil exploration and development contacts with many other countries and business entities through normal commercial channels. Nevertheless most Chinese authorities understand that the use of any fossil fuel will be only a tiny part of the time within which China has had a highly developed civilization. Hence China has begun the inevitable transformation to a society run on renewable energy, if that is possible.

Chapter 3
Possible Trends of Chinese Oil Supply Through 2030

In China, there are many who know that while oil production has increased steadily for several decades that oil is a finite substance whose production at some point cannot continue to increase. When will that time come? There have been a number of attempts to predict that as is developed in the next section.

3.1 Forecast of Chinese Oil Production

3.1.1 Weng Model

There's a natural process of "rise-growth-peak-decline" cycle in the life of almost anything, and the same goes for oil and gas discoveries and, subsequently, production (Hubbert 1969). On the basis of this theory, the deceased academician Wenbo Weng of the Chinese Academy of Science proposed the Poisson Cycle Model in 1984, generally called the "Weng Model," which was the earliest oil forecasting model in China, and this has laid the foundation for forecasting oil and gas production in China (Weng 1984). The basic formula for the Weng Model is:

$$Q = At^n e^{-t}, \ t \rangle 0 \tag{3.1}$$

Here, Q refers to annual production; t refers to development time, in years; A, n refers to a statistical parameter to fit the model to data. Essentially the first term (At^n) grows the production until the increased passage of time (t) in the second negative term (e^{-t}) becomes large enough to become more important. These terms reflect the development of knowledge about exploiting a resource and then the inevitably greater importance of depletion.

L. Feng et al., *The Chinese Oil Industry: History and Future*,
SpringerBriefs in Energy, DOI 10.1007/978-1-4419-9410-3_3,
© Lianyong Feng, Yan Hu, Charles A.S. Hall, Jianliang Wang 2013

3.1.2 Generalized Weng Model

In 1996, Professor Chen Yuanqian completed the theoretical derivation of the Weng Model. He was the first to propose the linear trial and error method for solving this non-linear model. The original Weng Model is a special case derived from the condition when the constant b is a positive integer, the modified model is named the "Generalized Weng Model" (Chen 1996).The important formulas contained in this model are as follows:

$$Q = at^b e^{-(t/c)} \tag{3.2}$$

$$Q_{max} = a(bc/2.718)^b \tag{3.3}$$

$$t_m = bc \tag{3.4}$$

$$N_R = ac^{b+1}\Gamma(b+1) \tag{3.5}$$

Here, Q refers to annual production; a, b, c are statistical parameters; 2.718 is e, the base of the natural logarithm, reflecting half the use of the resource; t refers to the time since the initiation of the development, a; Q_{max} refers to peak production; t_m refers to the year corresponding to the peak; N_R refers to Ultimate Recoverable Reserves (URR); $\Gamma(b+1)$ is the gamma function, when b is a positive integer, $\Gamma(b+1) = b!$.

3.1.3 Weibull Model

In 1995, Yuanqian Chen and Jianguo Hu completed a theoretical derivation by using the Weibull Distribution Model from mathematical statistics, and established the Weibull Forecasting Model (Chen and Hu 1995). The main formulas for the Weibull Forecasting Model are:

$$Q = at^b e^{-(t^{b+1}/c)} \tag{3.6}$$

$$Q_{max} = a\left[\frac{bc}{2.718(b+1)}\right]^{b/(b+1)} \tag{3.7}$$

$$t_m = \left(\frac{bc}{b+1}\right)^{1/(b+1)} \tag{3.8}$$

$$N_p = N_R[1 - e^{-(t^{b+1}/c)}] \tag{3.9}$$

$$N_{\mathrm{pm}} = 0.3679 N_{\mathrm{R}} \tag{3.10}$$

$$N_{\mathrm{R}} = \frac{ac}{b+1} \tag{3.11}$$

Here, Q refers to annual production; t refers to development time; a, b, c refer to statistical parameters; $Q_{\max}$ refers to peak production; t_{m} refers to the year corresponding to the peak; N_{p} refers to cumulative production; N_{R} refers to URR, the ultimate recoverable resource; N_{pm} refers to cumulative production corresponding to the peak.

The Weibull Model applies to oil and gas fields entering the decline period, which occurs when 37 % of the recoverable reserves have been extracted. This model does not belong to the category of symmetric distribution models.

3.1.4   HCZ Model

On the basis of statistical analysis of huge amounts of oil and gas fields' development data, and theoretical derivation, Jianguo Hu and Yuanqian Chen (1995a) proposed the HCZ Model. The main formulas for the HCZ Model are as follows:

$$Q = a N_{\mathrm{R}} e^{[-(a/b)e^{-bt}-bt]} \tag{3.12}$$

$$Q_{\max} = 0.3679 b N_{\mathrm{R}} \tag{3.13}$$

$$N_{\mathrm{p}} = N_{\mathrm{R}} e^{[-(a/b)e^{-bt}]} \tag{3.14}$$

$$t_{\mathrm{m}} = \ln \frac{(a/b)}{b} \tag{3.15}$$

$$N_{\mathrm{pm}} = 0.3679 N_{R} \tag{3.16}$$

$$\frac{Q_{\max}}{M_{\mathrm{pm}}} = b \tag{3.17}$$

Here, N_{p} refers to cumulative production; N_{R} refers to URR; a, b, c refer to statistical parameters; t refers to development time, a; Q refers to annual production; t_{m} refers to the year corresponding to the peak; $Q_{\max}$ refers to peak production; N_{pm} refers to cumulative production corresponding to the peak.

The HCZ Model applies to oil and gas fields entering the decline period, which occurs when 37 % of the recoverable reserves have been extracted, which is the same as with the Weibull Model. However, in regards to production in the declining stage, the HCZ Model has an obviously slower declining rate than the Weibull Model.

3.1.5 HC Model

Based on huge amounts of data on the development of oilfields, Jianguo Hu and Yuanqian Chen (1997) established the Hu-Chen Model. The formulas for the HC Model are as follows:

$$N_p = \frac{N_R}{(1 + at^{-b})} \tag{3.18}$$

$$Q = \frac{abN_R t^{-(b+1)}}{(1 + at^{-b})^2} \tag{3.19}$$

$$t_m = \left[\frac{a(b-1)}{b+1}\right]^{1/b} \tag{3.20}$$

$$Q_{max} = \frac{N_R(b-1)^{(b-1)/b}(b+1)^{(b+1)/b}}{4ba^{1/b}} \tag{3.21}$$

$$N_{pm} = \frac{(b-1)N_R}{2b} \tag{3.22}$$

Here, N_p refers to cumulative production; N_R refers to URR; a, b, c refer to statistical parameters; t refers to development time, unit: year; Q refers to annual production; t_m refers to the year corresponding to the peak; Q_{max} refers to peak production; N_{pm} refers to cumulative production corresponding to the peak.

3.1.6 Lognormal Distribution Model

Based on the lognormal distribution in mathematical statistics, Yuanqian Chen, and Zixue Yuan (1997) completed a model derivation. The formulas for the Lognormal Distribution Model are as follows:

$$Q = at^{-1}e^{1(\ln t - c)^2/b} \tag{3.23}$$

$$a = \frac{N_R}{\sqrt{\pi b}} \tag{3.24}$$

$$Q_{max} = ae^{(b/4-c)} \tag{3.25}$$

$$t_m = e^{(c-b/2)} \tag{3.26}$$

$$N_p = \int_0^t Q \mathrm{d}t \tag{3.27}$$

Here, Q refers to annual production; t refers to development time, a; a, b, c refer to statistical parameters; N_R refers to URR; Q_{max} refers to peak production; t_m refers to the year corresponding to the peak; N_p refers to cumulative production.

3.1.7 t Model

This model was proposed by Fusheng Huang et al. (1987). Jianguo Hu and Yuanqian Chen (1995b) completed the derivation of the model. The formulas for the t Model are as follows:

$$N_p = Ae^{\left(\frac{K}{n+1}t^{n+1}\right)} \tag{3.28}$$

$$Q_o = KAe^{\left(\frac{K}{n+1}t^{n+1}\right)}t^n \tag{3.29}$$

Here, N_p refers to accumulated cumulative production; A refers to URR; K, n refers to statistical parameters; t refers to development time; Q_o refers to annual production. This model only applies to the forecast of oil and gas reserves in the declining stage.

3.1.8 Generalized and Unified Oil and Gas Production Forecasting Model

Xinghua Yang et al. (2001) proposed the Generalized and Unified Oil and Gas Production Forecasting Model. The formula for this model is as follows:

$$Q = aN_p^k t^b \exp[-c_1 t^{m_1} - n\ln^p(c_2 t^{m_2} + c_3)] \tag{3.30}$$

Here, Q refers to annual production; a, k, b, $m1$, $m2$, c_1, c_2, c_3, n, p refer to statistical parameters; N_p refers to accumulated cumulative production.

If one has estimates of the model parameters from various oil fields, then various current production forecasting models can be derived. The formula covers all current models that forecast production, and possesses the basic characteristics of generalized models. The decreasing function term, which is $\exp[-c_1 t^{m_1} - n\ln^p(c_2 t^{m_2} + c_3)]$, essentially reflects the declining pattern of production inherent of oil fields in cases of different drive types and percolation characteristics, while the increasing function term, which is $N_p^k t^b$, reflects the production pattern under human influence. Different values are chosen for various parameters based on various drive types and percolation characteristics. After mathematical derivation and model simplification, various production forecasts can be obtained using this model. It is important to note that most of these models, when forced to use real data, are constrained by the data and so tend to give somewhat similar results for China as a whole, indicating a peak from 2009 to 2020 and a subsequent decline.

3.1.9 Multi-cycle Model

At present, most domestic scholars use the Single-Cycle Model for the forecasting of oil and gas production, including the previously mentioned Weng Model, HCZ Model, etc. However, there are certain limitations in these models when forecasting is conducted for the actual production of oil and gas fields. These limitations are mainly on of the following forms:

1. When the oil and gas fields have multi-cycle production cycles, there is a lower degree of fitting between the forecasted and historical production than when using Single-Cycle Models.
2. There is less accuracy in forecasting ultimate recoverable reserves when using the Single-Cycle Forecast Model for fields that are in fact characterized by multi-cycle production patterns. The Single-Cycle Model has a low degree of fitting degree with intermediate cycles other than the final cycle. The result is an increased discrepancy between the forecasted and historical production curve, thus different integral areas occur, leading to increased error of the forecasted ultimate recoverable reserves.

Since the Generalized Weng Model is appropriate for forecasting oil and gas production, Wang et al. (2011) have modified the Generalized Weng Model and established the Generalized Multi-Cycle Weng Model. The formulas for this model are as follows:

$$a = \frac{Q_{\max}}{(bc/e)^b} \tag{3.31}$$

$$c = \frac{t_m}{b} \tag{3.32}$$

$$Q = Q_{\mathrm{mas}} \left(\frac{et}{t_m}\right)^b e^{-\left(\frac{bt}{t_m}\right)} \tag{3.33}$$

$$N_R = \frac{Q_{\max}}{(t_m/e)^b} c^{b+1} \Gamma(b+1) \tag{3.34}$$

In the case of oil fields with k production cycles, the forecasted production can be obtained using the following formula (Feng et al. 2010):

$$Q(t) = \sum_{i=1}^{k} Q(t)_i = \sum_{i=1}^{k} \left\{ Q_{max} \left(\frac{et}{t_m} \right)^b e^{-\left(\frac{bt}{t_m} \right)} \right\}_i \qquad (3.35)$$

In the same manner, we can forecast the ultimate recoverable reserves using the following formula:

$$N_R(t) = \sum_{i=1}^{k} N_R(t)_i = \sum_{i=1}^{k} \left\{ \frac{Q_{max}}{(t_m/e)^b} c^{b+1} \Gamma(b+1) \right\}_i \qquad (3.36)$$

The Single-Cycle Model is a special case when k equals 1 in the Multi-Cycle Model.

3.2   The Forecast of Ultimate Recoverable Reserves and Oil Production in China

There is a close relation between the ratio of cumulative production to ultimate reserves (R_D) and development time (t), meaning the lapse between the first year of production and the present. Research on a huge amount of development data from oil and gas fields shows the relation between R_D and t using the following formula (Weng 1984; Zhao 1987; Huang et al. 1987):

$$\lg \frac{R_D}{1 - R_D} = A + B \lg t \qquad (3.37)$$

Here, A and B are coefficients, which we estimate through linear regression. The production degree of recoverable reserves (R_D) is determined by using the cumulative production (N_p) and URR in the following formula:

$$R_D = \frac{N_p}{U_{RR}} \qquad (3.38)$$

After a series of derivations of the previous formulas, we can obtain the ultimate recoverable reserves forecasting model, the process is as follows:

$$\frac{R_D}{1 - R_D} = 10^A t^B \qquad (3.39)$$

$$R_D = \frac{1}{1 + 10^{-A}t^{-B}}$$

(3.40)

Cumulative production (N_p) is obtained using the following formulas:

$$N_P = \frac{U_{RR}}{1 + 10^{-A}t^{-B}}$$

(3.41)

and,

$$N_p = \frac{1}{\frac{1}{U_{RR}} + \frac{10^{-A}}{U_{RR}t^{-B}}}$$

(3.42)

Since oil production is a first-order derivative of cumulative production, in order to derive two sides of the above equation we use the following formula:

$$Q = \frac{B10^{-A}U_{RR}t^{-(B+1)}}{(1 + 10^{-A}t^{-B})^2}$$

(3.43)

Then by using the two equations above, we can get:

$$\frac{Q}{N_P^2} = \frac{B10^{-A}}{U_{RR}}t^{-(B+1)}$$

(3.44)

$$\lg \frac{Q}{N_P^2} = \lg \frac{B10^{-A}}{U_{RR}} - (B+1)\lg t$$

(3.45)

Assuming α and β

$$\alpha = \lg \frac{B10^{-A}}{U_{RR}}$$

(3.46)

$$\beta = -(B+1)$$

(3.47)

Then, we get:

$$\lg \frac{Q}{N_P^2} = \alpha + \beta \lg t$$

(3.48)

Since N_P^2 and t are known data, we can find α and β using the following regression methods,

$$B = -(\beta + 1)$$

(3.49)

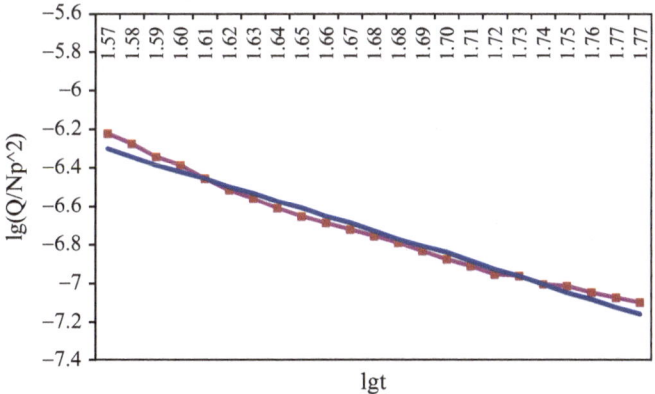

Fig. 3.1 Linear relation between $\lg(Q/N_p^2)$ and $\lg t$ from 1985 to 2007

$$\frac{10^{-A}}{U_{\mathrm{RR}}} = \frac{10^{\alpha}}{B} = -\frac{10^{\alpha}}{\beta + 1} \tag{3.50}$$

Therefore,

$$N_{\mathrm{P}} = \frac{1}{\frac{1}{U_{\mathrm{RR}}} + \frac{10^{-A}}{U_{\mathrm{RR}}} t^{-B}} = \frac{1}{\frac{1}{U_{\mathrm{RR}}} - \frac{10^{\alpha}}{\beta+1} t^{-(\beta+1)}} \tag{3.51}$$

Finally, the formula for the URR is:

$$U_{\mathrm{RR}} = \left[\frac{1}{N_{\mathrm{p}}} + \frac{1}{\beta + 1} 10^{\alpha} t^{\beta+1}\right]^{-1} \tag{3.52}$$

Using this formula, we can make a forecast of the ultimate recoverable reserves for China's petroleum.

There is a linear relation between $\lg Q/N_p{}^2$ and $\lg t$ from 1985 to 2007 in the Chinese oil industry (Fig. 3.1).

The ultimate quantity of oil that can be extracted in China (URR) can be estimated from this equation for each year, although the values obtained depend on the precise form of the equation and the values of the parameters. As a result, there are different forecasted results of China's ultimate recoverable oil reserves in different years. Although forecasted results of the URR are different each year, the gap is small and the different results are given in Table 3.1. We determined that the ultimate recoverable reserves are an average of 13.4 billion tons, which serves as a resource basis for the oil production forecast of China. This can be compared to the cumulative extraction from 1949 to date of 5.5 billion tons. This implies that China has used about 41 percent of its total recoverable oil resources.

Table 3.1 Forecasted ultimate recoverable reserves of China	Maximum	Minimum	Average
URR (billion tons)	14.3	12.3	13.4

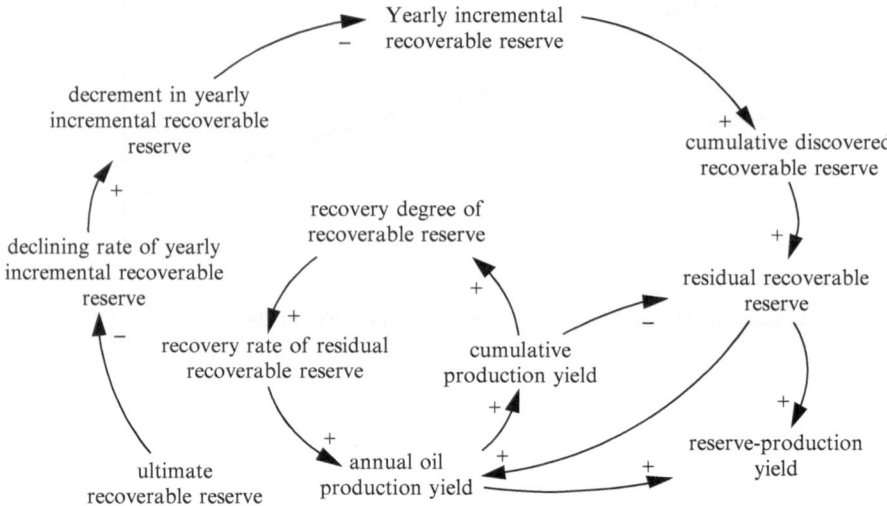

Fig. 3.2 Causal loop diagram of Chinese oil production forecasting

In most years before 1993, Chinese oil consumption was lower than production, in other words China was a net exporter of oil. After 1993, however, China became a net importer of oil with an increasing volume imported in successive years due to the rapid increase in oil consumption. Through sorting and analyzing exploration and development data in recent years in China, we established a causal loop diagram of Chinese oil production forecasting, based on system dynamics. In this model, there are 13 variables including: yearly incremental recoverable reserves, cumulative recoverable reserves, cumulative residual recoverable reserves, cumulative production, production degree of recoverable reserves, recovery rate of residual recoverable reserves, annual oil production and reserve-production ratio, etc. (Fig. 3.2).

It appears from this figure that the recoverable reserves and the recovery rate of remaining recoverable reserves are two important determinants in predicting annual oil production. The product after multiplication of these two variables is the annual oil production and the recovery rate of residual recoverable reserves. The latter is influenced by the degree of recovery of the remaining recoverable reserves. There is an increasing recovery rate of residual recoverable reserves along with the gradual increase in the recovery degree of the recoverable reserves. On the basis of principles of system dynamics, we established a System Dynamics Flow Diagram from the above casual loop diagram (Fig. 3.3).

For model-facilitating purposes, there are two ways to think about yearly incremental recoverable reserves, stocks and flows respectively. They are equal in value although they differ in nature.

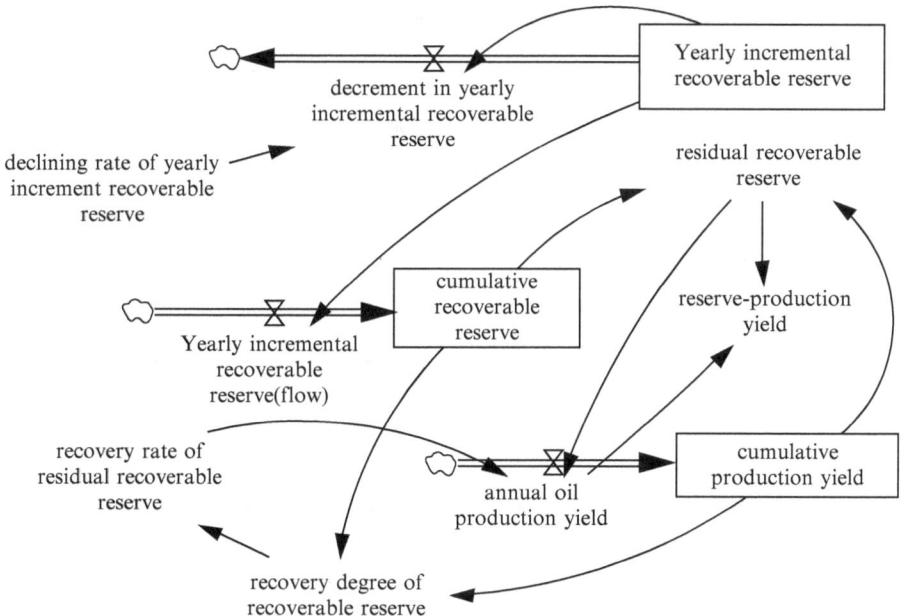

Fig. 3.3 System dynamics flow diagram of Chinese oil production forecast

We determined the initial value of certain variables in the model based on relevant historical data, e.g., annual cumulative recoverable reserves, residual recoverable reserves, cumulative production, yearly incremental recoverable reserves, etc. In the future, the balance between the older Eastern oil fields, the newer Northwest oil fields, and offshore oil fields will determine the future trends in the incremental recoverable reserves of China. We believe that Chinese yearly incremental recoverable reserves will maintain the current levels until about 2011. Then the Chinese yearly incremental recoverable reserves will decline at an average empirical rate of about 2.6 % after 2011.

The recovery rate of residual recoverable reserves is determined from the regression between the recovery rate of oil and the degree of recovery. The results show that there is a positive correlation between the recovery rate of Chinese oil and the degree of recovery.

Based on the previously established system dynamics model and relevant model parameters, we simulated yield of Chinese oil production until 2050. The forecasting results are given in Fig. 3.4.

From Fig. 3.4, it appears that (at least from our model results) that the current Chinese oil production is already in a "peak plateau" period, with not much room for future oil production growth. The forecasting results show that China will reach its peak oil production around 2015, with a production of around 192 million tons. After 2015, oil production will decrease continuously, falling below 100 million tons by 2050.

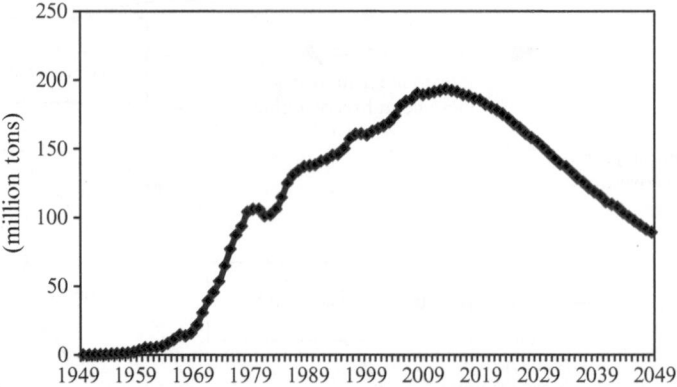

Fig. 3.4 Our forecast of Chinese oil production

3.3 Analyzing Chinese Oil Imports based on the Export Capacity of the World's Oil Producing Nations

3.3.1 The Importance of Analyzing Chinese Oil Imports

The rapid growth of China's economy requires enormous oil resources. Currently, both China's oil imports and degree of petroleum dependence are increasing rapidly. Because of the domestic oil production shortage, China has become more and more dependent on foreign oil. In 2009, China imported 218 million tons of oil and the degree of import dependence reached 53.5 %. If China cannot import enough oil in the future, the oil shortage problem will impact the development of industry and the national economy negatively.

3.3.2 Oil Production Forecasting Model and Oil Consumption Model

Our research procedures are (1) to select the major oil exporters, (2) to forecast and analyze their probable oil production and consumption into the future. The only difference is the export capacity for each country. Furthermore, we can estimate the world's oil export capacity by adding up each of these countries' export potential.

The historical data of production and consumption is from the BP statistical review of world energy (2009) and the EIA. The Chinese population data, both historical and projected, is from the United Nations. We used life cycle in oil and gas fields to forecast oil production.

The Oil Consumption per Capita (OCPC) method is used to forecast future oil consumption. First, forecasting the future OCPC according to the historical OCPC;

Table 3.2 Production of the five largest oil fields in Saudi Arabia

Oil field	Type	Discovery date (year)	Peak time(year)	Peak production (mbbl/day)
Ghawar	Onshore	1948	1980	5.59
Safaniyah	Onshore/offshore	1951	1998	2.13
Qatif	Onshore/offshore	1945	2006	0.50
Shaybah	Onshore	1968	2003	0.52
Zuluf	Offshore	1965	1981	0.68

Source: IHS, Deloitte & Touche, Database of USGS, IEA estimated and analysis

then, using the future Chinese population projections from the UN to predict the future oil consumption for each country. The basic oil consumption forecast formula is:

$$C_{for}(t) = P_{for}(t) \times OCPC_{for}(t) \tag{3.53}$$

where, C_{for}—oil consumption for year t, million tons; P_{for}—population for year t, million; $OCPC_{for}$—oil consumption per capita for year t, tons per capita.

3.3.3 Forecast of the World's Oil Export Capacity

Based on oil reserves and historical oil export data, we chose the 16 major oil exporters to analyze the world's oil export capacity: Saudi Arabia, Russia, the United Arab Emirates (UAE), Norway, Iran, Kuwait, Nigeria, Venezuela, Algeria, Angola, Libya, Iraq, Mexico, Kazakhstan, Canada and Qatar. Of these countries, 11 countries belong to OPEC. These 16 countries were responsible for more than 70 % of the world's oil exports in 2008, although their future capacity is somewhat in doubt as many are facing their own peak and all are consuming more of their oil internally (Hallock et al. 2004).

Saudi Arabia has the most abundant conventional oil resources and is the largest oil producer and exporter in the world. In 2008, Saudi Arabia had an official 36.3 billion tons of oil reserves, 515 million tons of oil production, and exported 411 million tons of oil. The corresponding proportions of world reserves, production and exports for Saudi Arabia were 21.0 %, 13.1 % and 15.2 %, respectively. There are more than 100 oil fields under production in Saudi Arabia, although 8 giant oil fields account for most of the production. All of these giant oil fields were discovered many decades ago and most of them are past their production peak (Table 3.2).

The production of major oil fields in Saudi Arabia will inevitably decline. Finding new giant oil fields will require a great deal of investment and the probability of success is very small. Even if these projects were implemented, the additional production from these projects is unlikely to offset the production declines in existing oil fields. These facts mean that it will be very difficult

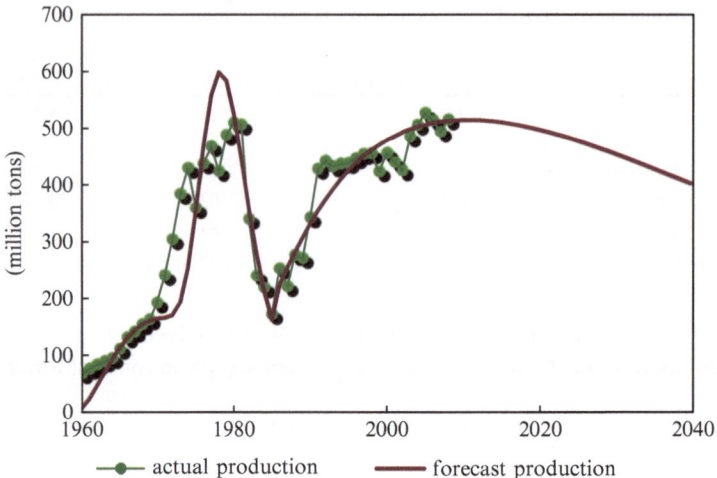

Fig. 3.5 Forecast results of oil production in Saudi Arabia

to increase or probably even maintain Saudi Arabia's oil production. From the forecast in Fig. 3.5, oil production probably will remain in a plateau for some years, and then decline. The peak in oil production will bring adverse impacts to the Saudi Arabian economy, which is almost entirely based on the oil industry. In order to maintain a sustainable level of development of its oil industry, King Abdul announced a prohibition on exploiting newly discovered oil fields in April of 2008. This indicates that Saudis have realized the impact of peak oil and are taking the necessary measures to reduce the impact on their economy in the near future.

Saudi Arabia is the largest oil consumer in the Middle East due to its rapid economic development. In 2008, Saudi oil consumption was 104 million tons, which represented 34% of the total Middle East consumption. From 2000 to 2008, the average growth rate of consumption has been 5.5 % per year. OCPC in Saudi Arabia began to increase since1993, and the average growth rate was 2.1 % per year from 1995 to 2006 (Fig. 3.6). Figure 3.6 shows that OCPC in Saudi will keep increasing steadily in the foreseeable future. However, its growth rate will slow down gradually because of a probable decrease in the rate of oil production. We predict that the average growth rate of oil consumption per capita in Saudi Arabia will be about 1.6 % per year from 2009 to 2020, and will shrink to about 1.4 % from 2020 to 2030 (Fig. 3.6).

We can predict Saudi Arabia's own oil consumption from its population increase and the basic per capita consumption forecast formula mentioned in the previous section. The UN Population Division forecasted Saudi Arabia's population in 2008. Assuming the medium growth rate, oil consumption in Saudi Arabia will continue to increase and reach 200 million tons in 2030 (Fig. 3.7). The average oil consumption growth rate is forecasted to be 1.68 % per year from 2010 to 2030.

Thus, the export capacity can be calculated as production minus consumption. This relation can be better understood using the following formula:

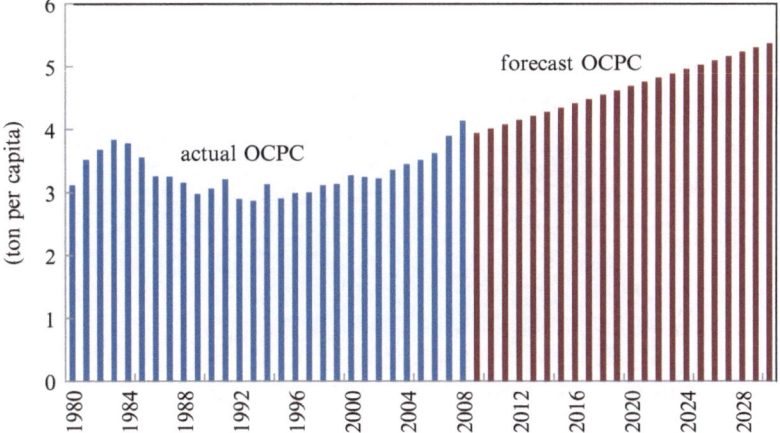

Fig. 3.6 Our results of oil consumption per capita (OCPC) in Saudi Arabia

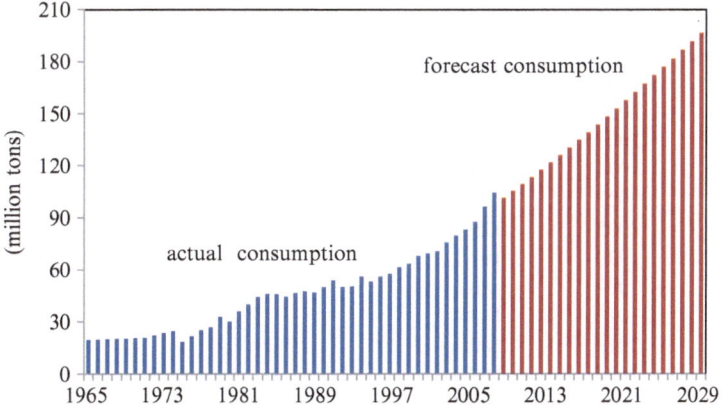

Fig. 3.7 Forecast results of oil consumption in Saudi Arabia

$$\text{Oil export (or import)} = \text{country's oil production}$$
$$- \text{country's oil consumption} \qquad (3.54)$$

Using this simple formula, we have calculated the oil export capacity of Saudi Arabia (Fig. 3.8). This figure shows that Saudi Arabian's oil export will decline in the future. Oil exports will be 350 million tons in 2020 and 260 million tons in 2030.

The oil export capacity of the remaining 15 major companies can be calculated following the same methods as Saudi Arabia. Each country's estimated oil production is shown in Table 3.3.

Many countries, including Norway, Venezuela and Mexico, and perhaps Saudi Arabia and Russia, have already passed their oil production peak. Angola, Libya,

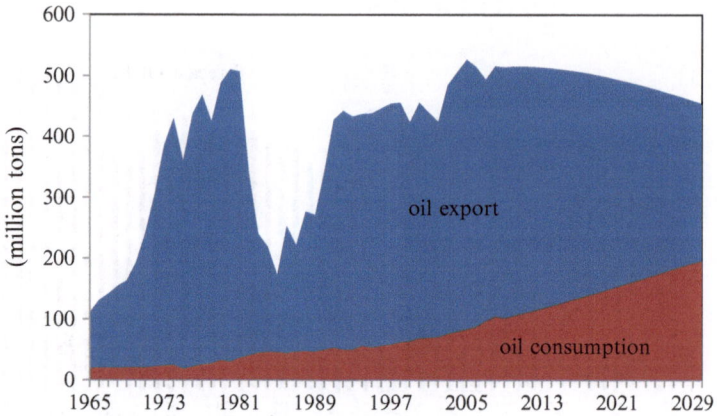

Fig. 3.8 Forecast result of oil export capacity in Saudi

Table 3.3 Oil production of the 15 major oil exporters

State	Peak time	Peak production (mbbl/day)	Production in 2020 (mbbl/day)
Saudi Arabia	2005	526.8	497.1
Russia	1987	569.5	427.5
UAE	2027	179.6	172.7
Norway	2001	162	36.6
Iran	1974	303.2	200.2
Kuwait	2027	168	164.2
Nigeria	2005	125.4	98.8
Venezuela	1970	197.2	104.8
Algeria	2007	86.5	58.5
Angola	2012	113.4	80.6
Libya	2018	100	99.7
Iraq	2027	167.4	162.1
Mexico	2004	190.7	79.1
Kazakhstan	2023	111.4	110.4
Canada	2054	224.2	179.5

Note: Oil production in Canada includes the oil sands. If oil sands aren't included, the peak of Canada's oil production will be in 2012, and peak production will be 161 million tons

Qatar, Kazakhstan, Iraq, Kuwait and UAE probably will peak between 2010 and 2030. Canada's oil peak may be very far away because of its abundant oil sands. Oil production of Kazakhstan and Canada will have some room for growth, but that growth is limited. Any expansion of production of oil sands in Canada needs substantial investment, which causes oil production in Canada to be full of uncertainty. The total oil production of the 5 non-OPEC countries is declining and almost certainly will decline in the future. Additionally, their oil export capacity will decline due to increased internal consumption caused by their economic growth, often fueled by cheap oil.

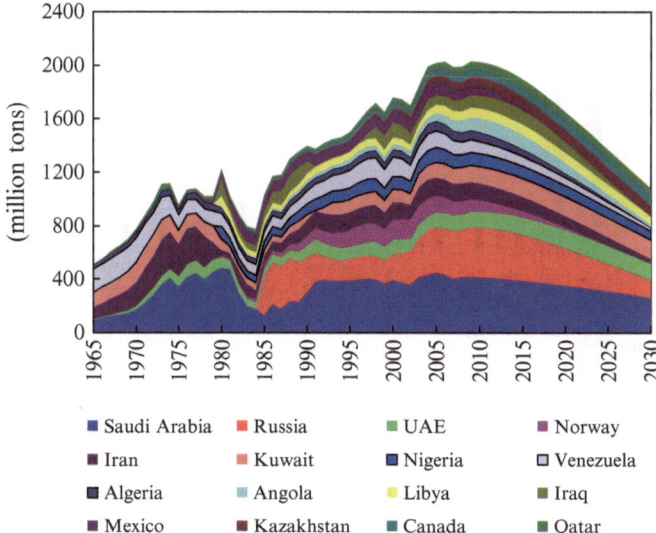

Fig. 3.9 Our model results of main oil exporters' export capacity

The oil production for the 11 OPEC countries fluctuates in part because of their use of oil as a "political weapon" to achieve their political purposes. Our forecasted results suggest that the total oil production of the 11 OPEC countries will continue to increase in the near future, and then it will begin to decline gradually over time. When compared with the non-OPEC countries, the peak oil time of the OPEC countries as a group is later.

Combining the OPEC and non-OPEC oil consumption forecasts generates the oil export capacity of these 16 countries (Fig. 3.9) which will peak and enter a plateau period in the near future, and then decline. We predict that the world's oil export capacity in 2010, 2020, and 2030 will be 2.03, 1.65 and 1.08 billion tons, respectively. The average decline rate is 2.02 % per year between 2010 and 2020, and 4.18 % per year between 2020 and 2030.

The oil export capacity for nine countries will decline before 2020, and for 15 of them it will decline by 2030 (Fig. 3.10). Only Canada has an increasing export capacity until 2030. According to the model results, Mexico, recently one of the world's export giants, will be a net oil importer soon, followed perhaps by Norway and Algeria between 2025 and 2030. From Fig. 3.10, we can see that Saudi Arabia, Russia, Kuwait and the UAE will probably be the four largest oil exporting countries in the future.

Previously, oil exports for the world have increased from 1984 until today. The countries noted above accounted for about 80 % of total world oil exports from 1994 to 2006. The percentage decreased to 73.7 % in 2008. If this percentage remains unchanged in the future, we can estimate the world's oil export capacity (Fig. 3.11). The peak plateau for the world's oil exports is between 2006 and 2016 at 2.75 billion tons, with this number declining to 2.24 billion tons in 2020 and 1.46 billion tons in 2030.

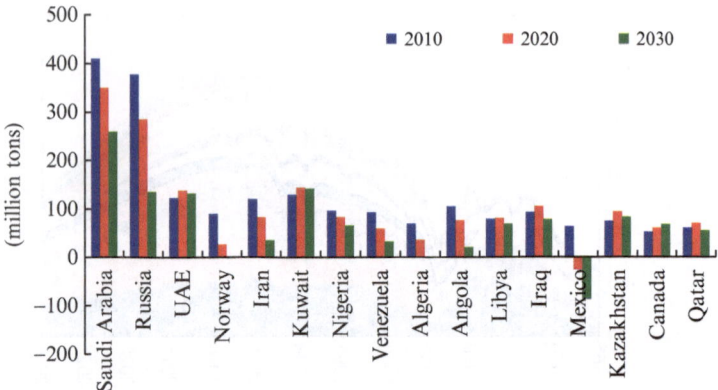

Fig. 3.10 Model results of main oil exporters' export capacity from 2010 to 2030

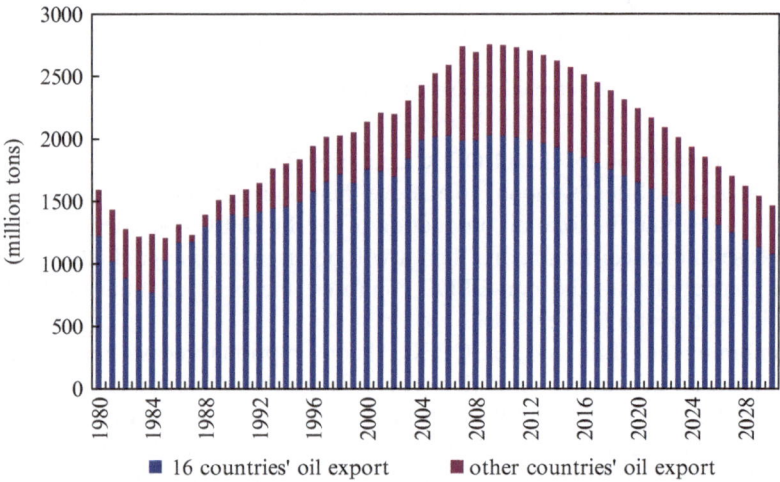

Fig. 3.11 World's oil export capacity outlook from 1980 to 2030. See Fig. 3.9 for countries involved

3.3.4 Forecast and Analysis of Chinese Oil Production and Consumption

China achieved oil self-sufficiency in 1965, and was a net oil exporter between 1973 and 1992. After 1993, China became a net oil importer because of rapid economic development with oil imports increasing year by year since then. Using the Multi-Cycle Generalized Weng Model to forecast future oil production (Fig 3.12), China's oil production will peak in 2015, and peak oil production will be less than 200 million tons. If the current economic development rate continues,

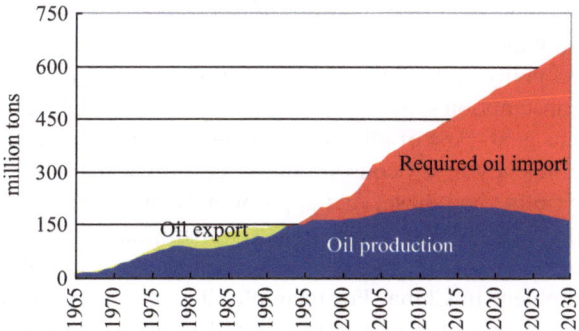

Fig. 3.12 Required oil import outlook for China

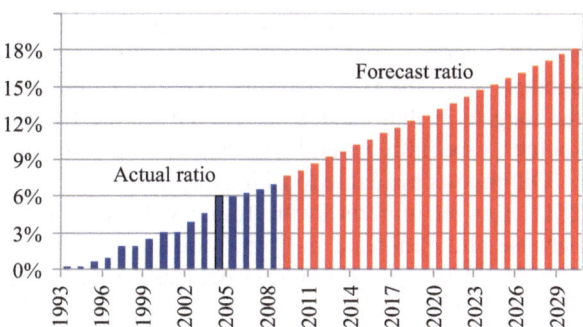

Fig. 3.13 The ratio of China's oil imports to the total oil exports of the world

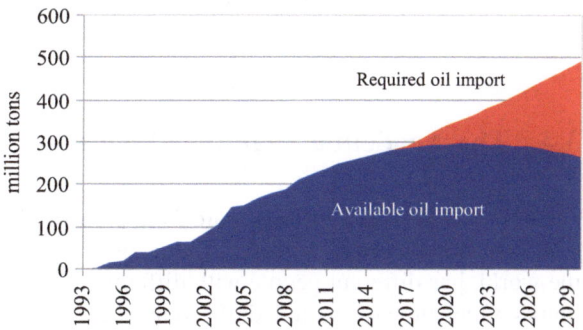

Fig. 3.14 Available world oil for import and required oil import for China

oil consumption in China will be about 530 million tons by 2020 and at least 650 million tons by 2030. This means that China will need to import 336 million tons of oil by 2020 and at least 488 million tons of oil by 2030.

Using the regression method, we can forecast the future oil ratio that China will import from total world oil exports in Fig. 3.13. It appears that available oil imports are less than China's projected demand after 2017 (Fig. 3.14), and the gap will grow and reach 43 million tons of oil in 2020 and 223 million tons of oil in 2030. This would be half of the oil available for export if the model represented by Fig. 3.9 is anything like correct.

Some effective measures must be taken to close the gap between available world oil imports and China's "required oil" imports to mitigate its negative effects on the economy. What can China do? First and foremost, we believe that excessive oil consumption growth must be curbed. We can take three possible actions to achieve this goal. First of all, energy savings might be an effective measure to improve the efficiency of oil consumption. Secondly, raising the price of oil products such as gasoline and diesel fuel appropriately may be another good measure to reduce the oil demand. Last but not least, transformation of the mode of economic development might be the most effective measure to solve the long term oil shortage problem in China. For example, it is the time for pertinent laws and regulations in China's automotive industry to be revised in order to encourage the development of energy efficient, low emission cars or to reduce the demand for cars entirely.

Developing alternative sources of energy, including wind energy, natural gas, nuclear energy, solar energy, geothermal energy and biomass energy, etc., is another very important measure to counter upcoming shortages of oil imports in the long term that cannot be ignored by the Chinese government. These energy sources, especially natural gas, can replace oil for many current applications.

Oil imports available to China may be able to meet the required oil imports until 2017. But oil imports in the future will become more and more difficult to get because of the increase in China's oil imports and assuming peaking world export capacity. This would require a decrease in consumption by other oil consumers. Such a situation would enormously exacerbate tensions between China and the rest of the oil-importing world. Therefore, the Chinese government and the experts must be aware of the seriousness of the problem, and adopt measures to solve or adapt to the problem of oil shortages in the future.

3.4 The Academic Peak Oil Debate in China

The above models and estimates are from peak oil theory, a perspective that we are very comfortable with. However, China, as is the case for different nations around the world, has different people with different opinions to debate on peak oil. Their points of debate have included: the year of estimated peak production, the roles of the technologies that are developing rapidly, the compensation to conventional fuels production by non-conventional and renewable energy (substitution) and the amount of fossil fuels reserves remaining under the ground etc. Hence, we have chosen to examine the basic positions of a number of people who have strong influences in China. They are divided into three groups: opposition, support, and neutrality with respect to peak oil.

Prior to discussing these three groups, we identified the first person to express his own opinions about peak oil. Bozhang Qian (2007) did not initially show his attitude clearly, on the one hand he accepted the concept of peak oil and on the other he accepted a high value on the history and potential of oil exploration and development. Even though he named his papers, "China Will Face Peak Oil in

2015" (Qian 2008) and "The Current Situation of Oil and Gas Production Peak in China" (Qian 2009), he did not agree or disagree with others, he just collected and put together other people's results.

The first group we will discuss is the opposition group to peak oil. Dadi Zhou, former leader of the Energy Bureau of NDRC, has discussed that the oil age has not disappeared at all and that we have not entered into the post oil era, but that we are only facing high oil prices (Wang 2009b). Those who would like to argue against peak oil believe that once there is peak oil, international oil prices will increase, not decrease. Besides that, they believe that any limitations to oil availability are due to monopoly and uneven share of power. In the Research Center of China's Company Go Global of the Ministry of Commerce of the People's Republic of China, Donghua Wu (2010) believes that the demand for oil will peak because of price increase, but the production of oil will not peak. She indicates that oil prices will begin decreasing and then that the theory of peak oil will be rendered ineffective. Kang Zhang (2008a, b, 2009), from Sinopec Research Institute of Petroleum Exploration and Development, fights against peak oil strongly, publishing three papers which give his arguments to prove peak oil wrong in many different ways. First, although he argues that the early forecasts by King Hubbert about the peak oil production of the lower 48 States of the United States was right, he also says that peak oil scientists avoid talking about Hubbert's incorrect prediction about the world's peak oil production. He argues that peak oil models are more suitable for simple, static and closed system such as countries, rather than a complex, dynamic and open system such as the whole world. Secondly, world oil production is not the same as one well or a small area; it has more and more new energy to continue producing oil. Wenrui Hu (2011), the former vice-CEO of Petro China Company Limited and chairman of China Oil Company Association, apparently does not disagree with the concept of peak oil, but he thinks that we can find more reserves if we keep exploring. He notes that world oil resources can be produced for more than 100 years at the current rate of exploitation and that we will not be faced with an oil supply shortage in the next 40–50 years. Thus, he states that we should think about oil and gas potential from a developing perspective. Senlin Hu (2011), who is an advanced leader in research policy of CNOOC, expressed his view that peak oil analyses ignore the scientific method in the background of the sophisticated world. If we put the peak oil concept into practice, it will face a lot of challenges. According to Hu, Hubbert's forecast in 1969 of global peak oil in 2000 is a false prediction because peak oil depends on reserves, technology, oil price, investment period, demand, and substitute energy. As a consequence, he believes that the concept of peak oil should be doubted.

Within the debate on peak oil, many researchers fall into groups of "agree" or "disagree" or "early" and "late" peak. However, there also is an indecisive group that not give its views about peak oil, but says something else which has a relation to it. We call the second group "neutrality." The former director of the energy bureau of NDRC, Dingming Xu (2009a, b), believes that, in the past 50 years, oil replaced coal and now renewable energy will replace oil to dominate in the future. He also believes that we will enter into a new age characterized by clean, high efficiency, low carbon and sustainable energy generation. Besides that, he believes that technology will decide the energy future and create future energy, and that human beings will shift from a society of energy resources to a society of energy technology. According to Xu, it will be a new energy revolution. The attitude from Qingyou Guan (2010), who is a Chinese economist, energy expert and distinguished researcher of GMEP, is that there is also peak consumption and peak emissions related to peak oil. The theory of these two peaks has increasing impact and can change the structure of world energy, world economy and politics. Boqiang Lin (Lin and Liu 2010), director of the China Energy Economic Research Center in Xiamen University, does not give an opinion about whether the research is wrong or right about peak oil, but he has forecasted that China's peak coal production will occur between the late 2020s and the early 2030s. He believes that peak oil theory is important for reserves, production and energy strategy, and national oil policy.

In China, a third group agrees with the existence of peak oil. This group includes ASPO-China, which was founded in 2005. It has being studying peak oil and supports the concept. So far, this group has aroused notice and attention from the public and officers by publishing more than 20 peak oil papers in journals and on the internet in China. This group is supported by Xiongqi Pang (Wang 2007a, b), the Vice-President of China's University of Petroleum. He believes that peak oil exists and that it will bring huge challenges to China, such as a severe shortage of petroleum supplies, increasing coal consumption, and environmental disruption. His views get attention from famous experts, but not from the level of important politicians and also not from the public who needs to understand this problem. Besides publishing papers, this group has visited some Chinese government and oil companies. However, they get less support there as this "limits to growth" perspective is not consistent with the government view of continued economic growth. Yuanqian Chen did not express any opinions about peak oil (including Qitai Yu (2002) and Jiye Wan (1994)), but his forecasts of China's oil resources show that oil production has peaked. He has also directed the research of ASPO-China for many years (ASPO-China 2005). Hongyi Hao et al. (2008), director of the policy research office at CNPC, believe that the world will have peak oil and international oil companies should join in this discussion. He believes that the oil companies' attitudes do not match their behaviors. They are taking some measures but they do not agree with the concept of peak oil despite their increasing difficulties in finding new oil supplies. Xianghong Cao (2008, 2009), the chief engineer in Sinopec, thought that the post-oil age is in sight and world oil production is near peak. If we are too optimistic about production, we will make a big mistake. Besides them, some experts from Chinese colleges and groups, such as the Business School of Hehai University (Tian 2008), Yangtze University (Li 2008), the School of Economics at Nankai University (Zhao and Zhao 2009; Zhao 2009a, b) and the Shanghai Alliance Investment Ltd. (Zhao 2009a, b), also have written papers to express their support of the concept of peak oil. However, they did not continue their study after they published their first papers. Here is a summary of opinions and backgrounds of influential people in China.

The loudest voices against peak oil are from several famous experts in China, and their backgrounds are almost always top level Chinese government officials and oil company leaders. Most experts who are unsure about peak oil are also from the top levels, and they often write in blogs to express their views and are often interviewed by the media. In comparison, Hongyi Hao and Xianghong Cao and ASPO-China clearly support peak oil theory. However, even though ASPO-China has lots of people who are students and teachers, they do not have much influence in China. Therefore, the voices that do not believe in peak oil have gained the most influence in China. This is the reason why ASPO-China has been hindered while promoting the peak oil theory.

This discussion of peak oil reflects similar discussions in the West, especially in the United States where peak oil deniers have been getting the most press, buoyed by some relatively small increases in US production (Hamilton 2012). More generally it reflects the tension between those who believe in some kind of "Limits to growth" and those who think it possible that human ingenuity and substitution will keep economies growing indefinitely. The original 1972 "Limits to growth" study (Meadows et al. 1972) has been almost universally attacked by economists and others, although several recent studies have shown that their basic model remains essentially almost exactly on course some nearly 40 years later (Hall and Day 2009; Turner 2008). While we believe that it is not possible to predict when or even that limits to growth will occur we believe it extremely foolish to not consider its possibility.

Chapter 4
Comprehensive Analysis of the Energy Return on Investment (EROI) of China

Whether or when peak oil occurs there is another, related issue that we think may be as or perhaps even more important, and certainly more general. This is a suite of specific concerns that surround the concept of Energy Return on Investment (EROI). We turn to this next.

4.1 Introduction of EROI

There are two general types of indices in the traditional analyses of economic systems: "gross" and "net." Most countries use them routinely in economic assessments such as the gross national product (GNP), their net national product (NNP), and as the gross income and net profit in financial analyses. Any economic system, from enterprises to countries, must consider not only their total input (i.e., revenue) but also the costs and the net output (i.e., profit) for their decision making. In the energy system, there is also a useful measure for net energy analysis, which is called Energy Return on Investment (EROI).

EROI is a method to calculate the energy returned to society compared to the energy required to obtain that energy. The units are usually dimensionless or may be in Joules per Joule, Calorie per Calorie, barrels per barrel etc. EROI evaluates the efficiency of energy production. Its value is one index of the net energy obtained rather than the gross energy, and of the eventual utility of the fuel to an economy. EROI is one measure of the value of an energy production system, which allows one to evaluate energy production physically rather than monetarily. It also appears to be an important determinant of price (King and Hall 2011). EROI is useful in two important ways: first for assessing the getting of energy itself, e.g., to help understand the normal process of oil exploration or development and production from an oil field or province or to compare different sources or fuels and second for the energy required to maintain and develop an economy and thus a society.

Although the concept is easy enough in theory, it is actually complex in practice. This has contributed to different results from different analyses of the same fossil fuels. Nearly all analyses show that the EROI for conventional oil, gas and coal is high but decreasing, and the EROI for alternatives including oil shale, liquefied coal and bio-fuels is much lower but in some cases increasing. To put it in perspective, in 1970 the oil and gas industry in the United States used the energy equivalent of one barrel of oil to produce about 30 barrels of oil, so the EROI is said to be about 30:1 (Cleveland et al. 1984). In the past decade, the EROI of oil for the United States has dropped to approximately 10:1, while the global EROI for oil generated by non governmental entities has declined from 35:1 in 1999 to 18:1 in 2006 (Gagnon et al. 2009).

EROI has an additional relevance to the concept of peak oil. Many economists, businessmen and some geologists are optimistic about the role of technology. They think that technological developments and capital investments can solve the peak oil problem. Thus, they may disagree, more or less, with the existence, timing or importance of peak oil. The truth is that there is race between depletion and technology, and there is no obvious way to distinguish which is the winner without empirical analysis of particular situations. Technology can indeed sustain or increase energy output, but it usually requires more substantial energy investments to do so. In a sense, the changes in EROI evaluate who is winning in the race between oil depletion and technological progress. It also shows the limits of the economists' argument that higher prices will bring in new supplies, for higher prices usually also mean higher energy use. That assessment by economists gives more credibility to peakoilism (Zhao et al. 2009).

4.2   Similar Indices Used in China

There are several existing indices of energy efficiency used in China whose equations are somewhat similar to that of EROI. These indices include "energy macro-efficiency," "productivity," "energy physical efficiency," and "efficiency of energy conversion." Just like EROI, each one of them evaluates the relations of "output" and "input." However, they are essentially different from EROI.

Energy intensity, one important index used to define "energy macro-efficiency" from the national point of view, uses GDP as the output and energy consumption as the input (4.1). A higher economic efficiency causes lower energy intensity. This energy index emphasizes the relation of energy consumption to the economy, and reflects the intricate dependence of the economy on energy at the national level. The energy use may or may not be corrected for energy quality, i.e., the economic utility of on Joule of electricity vs one Joule of coal.

$$\text{Energy intensity} = \text{energy consumption}/\text{GDP} \qquad (4.1)$$

Productivity uses the ratio of outputs and energy inputs to evaluate the production capacity from a macro-view perspective (4.2). Output is the gross (or net) national production measured in Yuan (or better in inflation-corrected Yuan), such as GDP. The input is based on human, material and financial resources or a combination of these resources.

$$\text{Productivity} = \text{commodities output/human labor (or capital or resources or other)} \tag{4.2}$$

Energy physical efficiency generally assesses energy consumption per unit product (4.3). This index focuses mainly on energy utilization per unit of energy-intensive products, such as total energy consumption per unit of steel production. It is especially suitable for the comparison between enterprises that have the same production structure, and the same level of equipment and management. There is a similarity between this index when applied to energy fuels and EROI, although they are reciprocal.

$$\text{Energy physical efficiency} = \text{energy consumption/production of product} \tag{4.3}$$

Efficiency of energy conversion is the ratio of energy output of a particular kind of energy to the energy inputs for processing and conversion of that product (4.4). This energy analysis focuses on the levels of equipment, and between equipment and the technology of conversion and management used to manufacture a product. In recent years, this energy index for petroleum refining and coking has been at about 95% and above. The same ratio for electricity power stations fluctuates at about 39%.

$$\text{Efficiency of energy conversion} = \text{energy output after conversion/} \\ \text{energy input for that conversion} \tag{4.4}$$

In China, most of these ratios are applied at the level of the entire economy or some other large entities.

EROI is different and perhaps more important than these energy indexes because it examines the effectiveness of obtaining energy itself, which in turn determines all the other ratios. EROI assesses the energy gain relative to energy costs, and assesses how the quality of the energy base is changing over time, including changes in net energy gains from energy resources. It is usually applied at the level of a particular field, region or political unit. More importantly, EROI allows comparisons between different energy fuels and provides a way to examine trends over time between the relation of technology and depletion. We believe that EROI should become one of the important components of China's official energy statistics and be added to the four energy indexes above. There may be less enthusiasm for governments to maintain such statistics because, unlike the other indexes, EROI often shows declines over time. This is in opposition to the concept of continual technological progress which the Chinese government likes to project. For that reason we are not optimistic that EROI will be added to the other four indexes used for regular energy calculations.

4.3   EROI Methodology

4.3.1   Formulas for EROI

In an attempt to formalize the analysis of EROI and to reduce or at least understand the discrepancies, Murphy et al. (2011) used a two-dimensional framework for EROI analysis: the horizontal dimension is "what do we count as energy output?" using three (or more) system boundaries, such as the mine mouth, refined fuel or including costs up to and including point of use. The vertical dimension is "what do we count as inputs?" The inputs and outputs can be in thermal units or quality-corrected. Thermal equivalents are based on the first law of energy conservation, i.e., that all energies can be measured by their final conversion to heat. The most common method of deriving EROI is to compare energy outputs and inputs in thermal units. The formula is:

$$\text{EROI} = \frac{E_o}{E_d + E_i + E_{lab} + E_{aux} + E_{env}} \tag{4.5}$$

E_o refers to the gross flow of all energy output for the period considered. The EROI protocol (Murphy et al. 2011) describes the energy inputs into the following five levels in the equation E_d, E_i, E_{lab}, E_{aux} and E_{env} meaning direct inputs, indirect inputs, labor consumption, auxiliary services consumption, and environmental costs, respectively. Since most studies to date have used both direct energy and indirect energy inputs, but not labor or environmental costs. Some energy costs are available only as monetary values, and need to be converted to energy. Thus the "standard" EROI formula typically uses the following equation:

$$\text{EROI}_{stnd} = \frac{E_o}{E_d + E_i} \tag{4.6}$$

or

$$\text{EROI}_{stnd} = \frac{E_o}{E_d + (M_i \times E_{ins})} \tag{4.7}$$

where E_o is joules of all energy outputs expressed in the same units, E_d and E_i represent direct (on site) inputs of different kinds of energy and indirect (off site) inputs of embodied energy, such as embodied in steel forms, cement, pipes and equipment (4.6). Because most of the time we cannot get all the indirect input energies directly, we give (4.7), where M_i expresses indirect inputs in monetary term and E_{ins} is the energy intensity of a dollar input for indirect components such as pumps or pipes.

Table 4.1 Conversion factors from physical units to thermal units

Energy	Average calorific value
Raw Coal	20.9 MJ/kg
Cleaned Coal	26.3 MJ/kg
Other Washed Coal	
Middling	8.36 MJ/kg
Slimes	8.36~12.5 MJ/kg
Coke	28.4 MJ/kg
Crude Oil	41.8 MJ/kg
Fuel Oil	41.8 MJ/kg
Gasoline	43.1 MJ/kg
Kerosene	43.1 MJ/kg
Diesel	42.7 MJ/kg
Liquefied Petroleum Gas	46.1 MJ/kg
Refined Gas	46.1 MJ/kg
Natural Gas	38.9 MJ/cu. m
Coke Oven Gas	16.7~18.0 MJ/cu.m
Other Coal Gas	
By Gas Furnace	5.23 MJ/cu.m
By Heavy Oil Catalytic Cracking	19.2 MJ/cu.m
By Heavy Oil Thermal Cracking	35.6 MJ/cu.m
Coke Gas	16.3 MJ/cu.m
By Pressure Gasification	15.1 MJ/cu.m
Water Coal Gas	10.5 MJ/cu.m
Coal Tar	33.5 MJ/kg
Electricity (in calorific value)	36.0 MJ/kWh

Different grades of fuels may vary considerably from the average. Datum is from China's Energy Statistical Yearbook
Source: China Energy Statistic Yearbook 2010

4.3.2   Energy Intensity

It is easy to convert physical energy units because they have fixed, if sometimes approximate, conversion factors (Table 4.1). Sometimes indirect inputs are given in physical units, and then energy costs can be derived fairly unambiguously. However, most of inputs are available only in monetary or currency units, and it is more difficult to derive the quantity of energy embodied in them. Those inputs, especially indirect inputs, usually require energy intensity (E_{ins}) to convert them into energy units. The energy intensity is the energy associated with the production of one monetary unit of a good or service and can be used as a proxy variable for the conversion of monetary values into joules.

In theory, energy intensity should be based on numbers from e.g., the entire (national) oil industry or related to the national oil and gas supply sectors—in other words, it should be for those sectors supplying the indirect inputs such as steel forms or drill bits. At one time this was quite possible to do, at least for the United States, because elaborate energy Leontief (i.e., I-O or input output tables

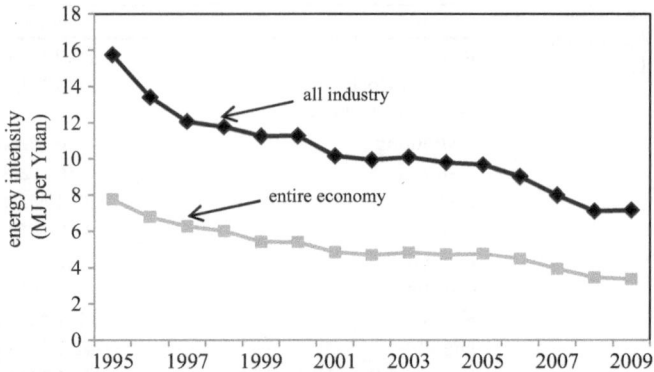

Fig. 4.1 Energy Intensity for: (1) all industry and (2) the entire economy in China.Data obtained from China's Energy Statistical Yearbook. We did not correct the economic data for inflation, so most of the apparent decline in energy intensity is due to inflation

had been generated by Investigators at the University of Illinois (e.g., Bullard and Herendeen 1975). Unfortunately such analyses have not been continued. It is difficult to calculate the energy intensity of specific entities in China because of the lack of data on value added and energy consumption for specific sectors of the economy. In the Chinese data, the sectors for extraction of petroleum and natural gas are categorized simply as belonging to "industry." Therefore, we must use the more general conversion factors for all industry, and alternately for the entire economy. We were able to get data for value added and energy use for all industry and for the entire economy in China to derive a time series of energy intensities (Fig. 4.1). We eliminate the effect of inflation by deriving the ratio of each year. The energy intensity for industry is about twice as that for the average of the entire economy which is similar to what has been found in other countries (e.g., US) (Murphy et al. 2011).

4.3.3 Correcting for Energy Quality

All energy sources have two properties: "quantity" and "quality." The thermal equivalent merely reflects "quantity," and does not represent energy "quality." Indeed, it is the different qualities among energy resources that cause the differences in their economic values even though they may have equivalent heat contents. Quality reflects work capacity, energy concentrations, scarcity security, cleanliness etc. (Zarnikau et al. 1996).

Kaufmann (1994) made an assumption that energy resources can complement each other completely, and market price reflected their qualities. In this case, the energy quality factor is expressed in the following formula:

$$\lambda_{it} = \frac{P_{it}}{P_{1t}} \tag{4.8}$$

where P_{1t} means the reference price in the time of t, P_{it} refers to the price of energy i at the same time. The type of reference energy and its price could impact the results greatly.

We used the Divisia index (4.9), which makes the assumption that the quality of a fuel is related to its price per heat unit (Berndt 1978, 1990). While price is not a perfect measure of energy quality, it is better than no correction at all. Besides, data on energy price is easier to get. Quality correction is the most important here with respect to the quality differences between oil, natural gas and other energy (Brown and Ulgiati 2004).

$$\ln E'_t - \ln E'_{t-1} = \sum_{n=1}^{k} \left(\left(\frac{P_{nt}E_{nt}}{2\sum_{n=1}^{k} P_{nt}E_{nt}} + \frac{P_{nt-1}E_{nt-1}}{2\sum_{n=1}^{k} P_{nt-1}E_{nt-1}} \right) (\ln E_{nt} - \ln E_{nt-1}) \right) \tag{4.9}$$

where p is the price of n different types of fuels and E is the final consumption of energy (joule) for each fuel type.

4.4 Application 1: EROI of China's Oil and Natural Gas Extraction

Oil and natural gas have become very important sources of energy for the Chinese economy. Assessing the EROI for Chinese oil and natural gas extraction becomes significant because it is becoming much more expensive to get them. We use energy outputs and inputs data from the China Statistic Yearbook which maintains average data for all of China's oil and gas extraction to calculate $EROI_{stnd}$. Annual data was available for crude oil production, natural gas production, and the energy inputs for getting those products available. We combined oil and natural gas production because the energy used as input was not different between oil and natural gas.

4.4.1 The Numerator: Energy Outputs

Data for the energy content of crude oil and natural gas production is given as tons coal equivalent (tce) for each year and as the percentage that is crude oil versus that of natural gas. We converted crude oil to tonnes and to joules, and natural gas production cubic meters and to joules (Table 4.2 and Fig. 4.2).

Table 4.2 Production of crude oil and natural gas in China

Year	Crude oil (raw data) (10^6 tce)	(10^6 ton)	(10^{18} J)	Natural gas (raw data) (10^6 tce)	(10^9 cu.m)	(10^{18} J)	Total energy outputs (10^{18} J)
1995	214	150	6.3	24.2	18.2	0.7	7.0
1996	225	157	6.6	26.7	20.1	0.8	7.4
1997	230	161	6.7	27.6	20.7	0.8	7.5
1998	230	161	6.7	28.3	21.2	0.8	7.6
1999	229	160	6.7	33.5	25.2	1.0	7.7
2000	233	163	6.8	36.2	27.2	1.1	7.9
2001	234	164	6.9	40.3	30.3	1.2	8.0
2002	239	167	7.0	43.4	32.6	1.3	8.3
2003	242	170	7.1	46.6	35.0	1.4	8.4
2004	251	176	7.4	55.1	41.4	1.6	9.0
2005	259	181	7.6	65.6	49.3	1.9	9.5
2006	263	184	7.7	77.9	58.6	2.3	10.0
2007	266	186	7.8	92.0	69.2	2.7	10.5
2008	276	193	8.1	106.6	80.2	3.1	11.2
2009	271	190	7.9	113.4	85.3	3.3	11.2
2010	291	204	8.5	125.3	94.2	3.7	12.2

Datum is from China Statistic Yearbook for each year. 1 kg crude oil = 1.4286 kg coal equivalent crude oil. 1 cu.m natural gas = 1.33 kg coal equivalent natural gas

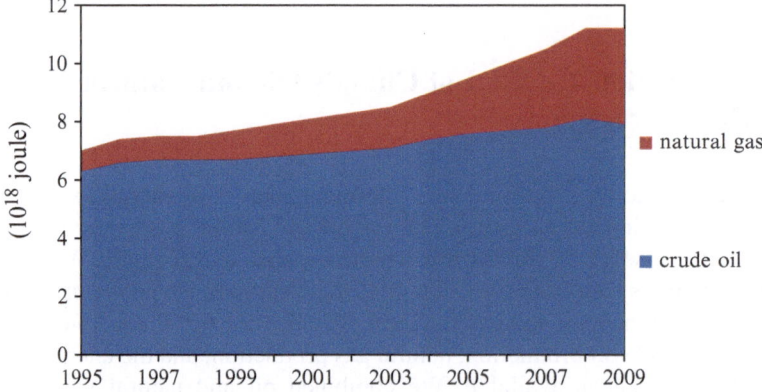

Fig. 4.2 Energy outputs of oil and gas extraction. *Source*: China Energy Statistic Yearbook 2010 and China Statistic Yearbook 2010

4.4.2   The Denominator: Energy Inputs

The Chinese Statistic Yearbook typically gives data for natural gas, crude oil, electricity, diesel oil, raw coal, fuel oil, gasoline, and refinery gas for oil and natural gas industry group in China. All raw data for direct inputs with physical units are given in Appendix 1. We converted all raw data into thermal units using the

Table 4.3 Direct energy inputs to the petroleum and natural gas sectors in China (unit: 10^{15} J, PJ)

Year	Natural gas	Crude oil	Electricity	Diesel oil	Raw coal	Fuel oil	Gasoline	Refinery gas	Total
1995	162	73	93	63	46	70	25	12	543
1996	117	71	93	63	47	33	13	3	440
1997	163	132	113	67	67	46	15	16	621
1998	151	131	107	45	45	54	14	20	568
1999	177	138	111	62	37	62	18	26	632
2000	196	171	116	69	40	63	20	28	702
2001	227	177	128	76	36	64	19	28	754
2002	231	187	131	84	34	61	19	29	777
2003	240	231	128	72	36	52	17	28	804
2004	189	209	131	79	36	14	16	17	689
2005	190	211	139	79	36	11	11	18	694
2006	212	236	114	80	37	12	13	20	724
2007	249	238	112	84	36	11	13	15	760
2008	336	291	115	116	30	16	12	17	933
2009	346	204	120	98	30	11	11	15	835
2010	398	202	125	79	34	14	10	16	878

conversions in Table 4.1. We then summed these direct inputs with thermal unit for the oil and natural gas extraction sector (Table 4.3).

Indirect inputs (off-site) (Table 4.4) are all of the energy used to generate the materials, equipment etc. required the on-site activities. Such indirect energy inputs, also called embodied energy, are usually in dollar cost of equipment or materials. We derived indirect energy inputs by multiplying costs by the energy intensities, because there is no energy accounting in routine economic data. Indirect monetary costs are from "purchase of equipment and instruments" and "other expenses" of "investment" in the item named as "investment in fixed assets in urban area by sector and total investment in construction" in China Statistical Yearbook database.

Total energy inputs were the sum of direct and indirect energy inputs (Table 4.5 and Fig. 4.3).

4.4.3   Results

We found that the energy return on energy investment for China's oil and natural gas extraction showed the slow declining trend from the maximum value at 14:1 in 1996 to 10:1 in 2010 with the annual decreasing rate of 2.6%. In recent years, the EROI has not over the maximum value even though it increased in 2009 and 2010. (Fig. 4.4).

$EROI_{stnd}$ declines when drilling effort increases in the United State as has they found by Hall and Cleveland (1981) and Guildford et al. (2012). In China, there was also an inverse correlation between $EROI_{stnd}$ and drilling efforts even though it was

Table 4.4 Indirect energy inputs to the petroleum and natural gas sectors in China

| | Raw data (10^9 Yuan) | | | Conversion | |
Year	Purchase of equipment and instrument[a]	Other expenses[b]	Total	Energy intensity for all industry (MJ/Yuan)	Total (PJ)
1995	4.5	2.2	6.7	11.3	75
1996	5.2	2.3	7.5	9.6	72
1997	5.5	2.6	8.1	8.7	70
1998	6.1	3.2	9.3	8.4	78
1999	6.1	3.0	9.1	8.2	75
2000	6.9	3.6	10.5	7.6	80
2001	7.7	4.2	11.9	7.2	86
2002	8.6	5.6	14.2	7.0	99
2003	11.0	7.5	18.5	7.0	129
2004	13.6	10.5	24.1	6.8	165
2005	19.8	12.9	32.8	6.4	210
2006	25.1	19.7	44.8	5.9	265
2007	25.9	16.5	42.4	5.3	225
2008	44.4	20.2	64.6	4.7	304
2009	55.2	21.6	76.8	4.7	364
2010	58.4	24.0	82.5	4.2	347

Note: [a] Purchase of equipment and instruments refers to the total value of equipment, tools, and instruments purchased or self-produced which come up to the cutoff point for fixed assets by the construction units or investing enterprises or institutions. Equipment, tools and instruments purchased or self-produced for new workshops by newly established or expanded units are categorized as "purchase of equipment and instruments" no matter whether they come up to the cutoff point for fixed assets
[b] Other expenses refer to expenses arising during the construction or purchase of fixed assets other than those mentioned in the above description

Table 4.5 Total energy inputs of the petroleum and natural gas sectors in China (unit: 10^{18} J (EJ))

Year	Direct energy inputs	Indirect energy inputs	Total energy inputs
1995	0.54	0.08	0.62
1996	0.44	0.07	0.51
1997	0.62	0.07	0.69
1998	0.57	0.08	0.65
1999	0.63	0.07	0.71
2000	0.70	0.08	0.78
2001	0.75	0.09	0.84
2002	0.78	0.10	0.88
2003	0.80	0.13	0.93
2004	0.69	0.17	0.85
2005	0.69	0.21	0.90
2006	0.72	0.27	0.99
2007	0.76	0.23	0.98
2008	0.93	0.30	1.24
2009	0.84	0.36	1.20
2010	0.88	0.35	1.23

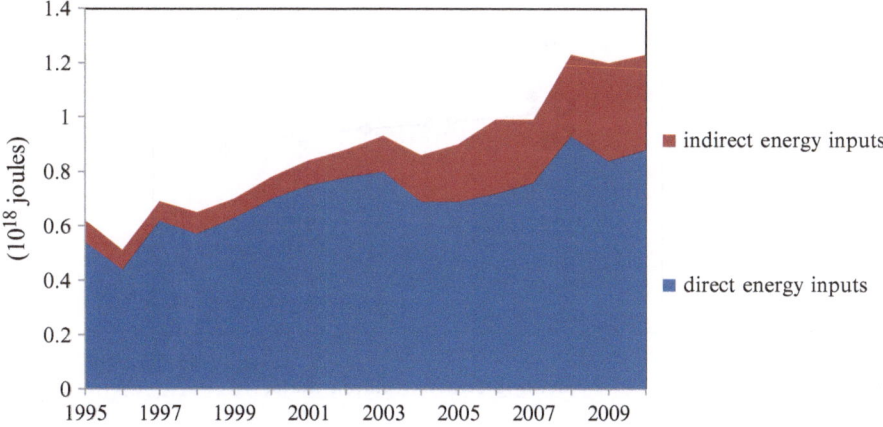

Fig. 4.3   Direct and indirect energy inputs

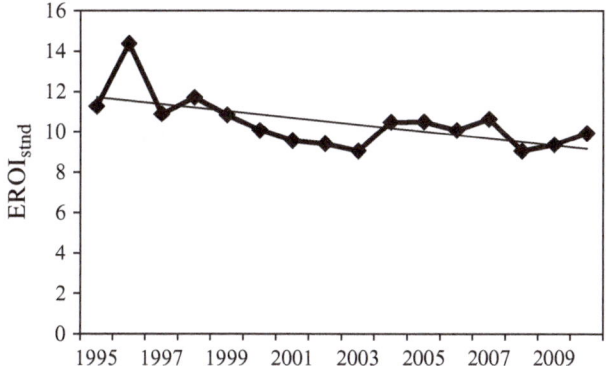

Fig. 4.4 $EROI_{stnd}$ of petroleum and natural gas sector in China

not too obvious (Fig. 4.5). $EROI_{stnd}$ tended to be higher with a little lower drilling footage. The increasing drilling efforts did not necessarily generate more oil and natural gas production but required more energy consumption.

4.4.3.1   Sensitivity Analysis

The energy intensity is not known with certainty. It associates each monetary unit spent for indirect cost with the energy used. The basic assumption is to use the energy intensity of all industries, but we examined the effect of different energy intensities for each yuan spent. We used different values of energy intensity of industry and the entire Chinese society in order to get different indirect energy inputs and $EROI_{stnd}$ for each year. The result suggests that energy intensity does not

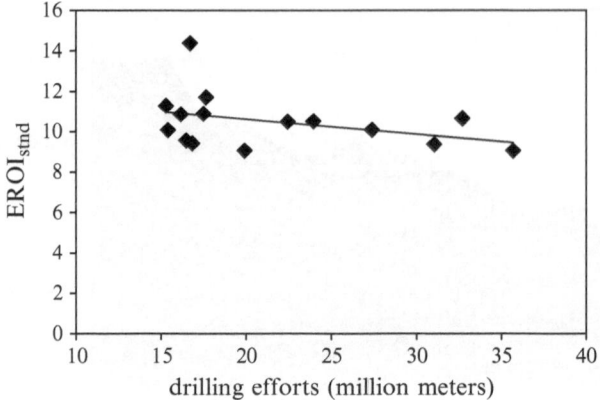

Fig. 4.5 EROI and drilling effort from 1995 to 2010 for China's oil and natural gas industry

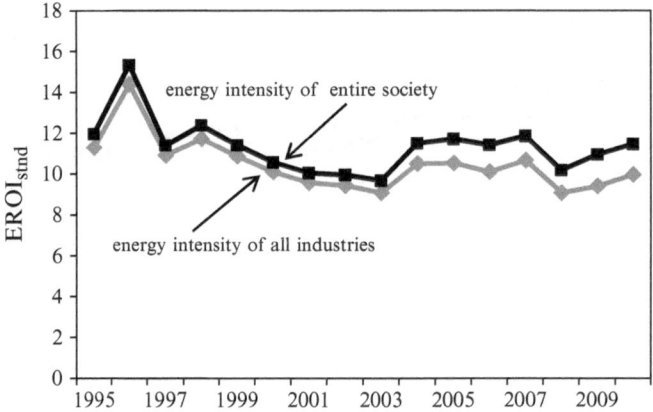

Fig. 4.6 EROI$_{stnd}$ with the energy intensity of the entire Chinese economy and all its industry

largely impact EROI$_{stnd}$ for China's oil and natural extraction. The energy intensity of entire society is about one of average larger than with that of all industries (Fig. 4.6), while the energy intensity of all industries is twice as large than that of the entire society (Fig. 4.1).

4.5 Application 2: EROI of China's Largest Oil Field-Daqing

We considered next the processes of exploration, development and production of oil and gas from the Daqing oil field, China's largest. We examined EROI using just direct and direct plus indirect energy inputs (total energy inputs). We also have

Table 4.6 Two-dimensional framework for EROI analysis

Level	Energy inputs	Heat equivalents	Quality-corrected heat equivalents
Level 1	direct energy inputs	$EROI_{1,d}$	$EROI1_{1,\ Qd}$
Level 2	indirect energy inputs	$EROI_{stnd}$	$EROI_{1,\ Qstnd}$

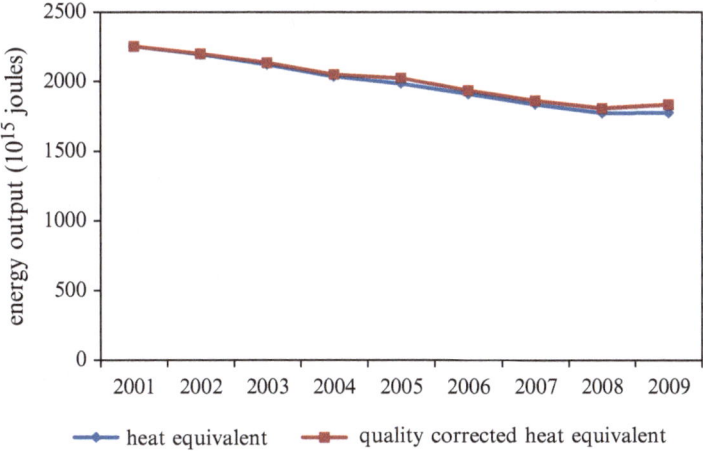

Fig. 4.7 Energy outputs expressed as heat equivalents and quality-corrected heat equivalents

made a quality correction for direct energy input in order to get the total energy inputs in quality-corrected heat equivalents (Table 4.6). In Table 4.6, the subscript "1" means the boundary for system boundary 1, while the "d" refers to direct energy inputs. $EROI_{stnd}$ represents the direct and indirect energy inputs and outputs from boundary 1. The "Q_d" refers to direct energy inputs in quality-corrected heat equivalents, while the "Q_{stnd}" refers to quality-corrected of "$EROI_{stnd}$."

4.5.1 The Numerator: Energy Outputs

The data of the Daqing oil field output was derived from the information available in the official web site for the field. This output was converted to heat units using the values in Table 4.1. We also converted all energy units to a common unit using the Divisia index to weigh the different energy qualities. To calculate the Divisia index, energy prices are the key factor. We were able to get accurate crude oil prices for the Daqing oil field every year (from trends on Oil Prices 2003–2006; Oil Price Trends 2007; Oil Price and Related Index Trends 2008–2009; Trend of Oil Price and Relevant Indexes 2010–2011). The NDRC has published adjusted gas prices several times over the course of nine years for different uses. We derived the annual average gas price for industries (NDRC 2011). Then, we got energy output corrected for quality (Fig. 4.7).

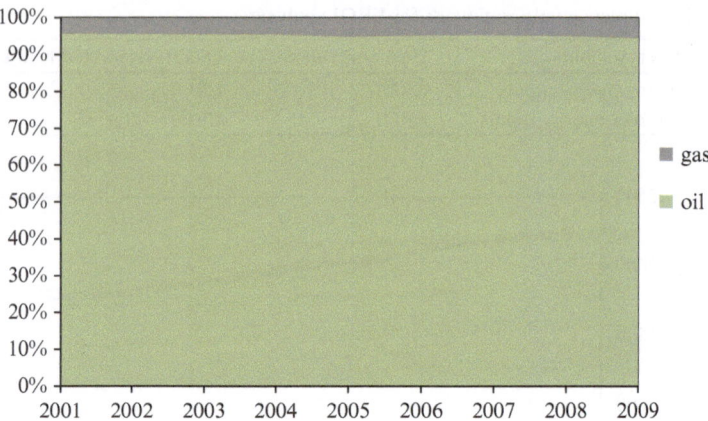

Fig. 4.8 Percentage of oil and gas joules in the total energy output. *Source*: China energy statistic yearbook 2010 and China statistic yearbook 2010

The difference between quality-corrected and non-corrected energy output is very small. There are two main reasons for this situation. The first is that natural gas production is small compared to oil production (4.4% in 2001, and 5.6% in 2009; Fig. 4.8). Thus the natural gas production expressed as joules gives only a very small correction. The second reason is that natural gas prices, which are lower than the true market prices, are controlled by the NDRC.

4.5.2 The Denominator: Energy Inputs

1. Direct Energy Inputs

   Direct input is given for the Daqing oil field in physical units (ton or kWh), and consists of oil for self-use, natural gas for self-use, gasoline, diesel, and electricity (Table 4.7). Some of the direct inputs, such as water, cannot be defined readily in energy terms, so we ignored them. The gasoline and diesel prices come from the pricing policy of NPRC, which we used to calculate annual average prices. Electricity price is based on 0.572 Yuan/kWh from 2001 to 2008 and 0.595 Yuan/kWh in 2009.The inputs were also expressed in quality-corrected terms using the same Divisia method as for energy outputs (Fig. 4.9). The maximum percentage difference in quality-corrected heat equivalents compared to heat equivalents is only 3.8%.

2. Indirect Energy Inputs

   No explicit data was available for indirect energy inputs for the Daqing oil field. However, we were able to estimate indirect cost using the following equation:

$$E_{\text{Indirect}} = (M_{\text{Total}} - M_{\text{Direct}}) \times E_{\text{ins}} \times C_{\text{CE–RC}} \times C_{\text{T–J}} \qquad (4.10)$$

Table 4.7 Money inputs to the Daqing oil field in 2002

Total Inputs (raw data)			Unit			As Money (10^3 Yuan)
operating costs			10^3 Yuan	11,390,080		11,390,080
depreciation			10^3 Yuan	10,625,160		10,625,160
expenses			10^3 Yuan	3,588,010		3,588,010
Total money inputs (M_{Total})						25,603,250
Direct Inputs (raw data)		Unit	Price	Unit		As Money (10^3 Yuan)
oil for self-use	200.0	10^3 t	1,558,012	Yuan/10^3 t		312,070
gas for self-use	1.2	10^9 m^3	920,000,000	Yuan/10^9 m^3		1,130,680
gasoline	36.3	10^3 t	303	Yuan/10^3 t		11.0
diesel	62.1	10^3 t	273	Yuan/10^3 t		16.9
electricity	9.8	10^9 kWh	571,700,000	Yuan/10^9 kWh		5,596,371
Direct money inputs (M_{Direct})						7,039,149
Indirect money inputs (($M_{Indirect} = M_{Total} - M_{Direct}$)						18,564,101

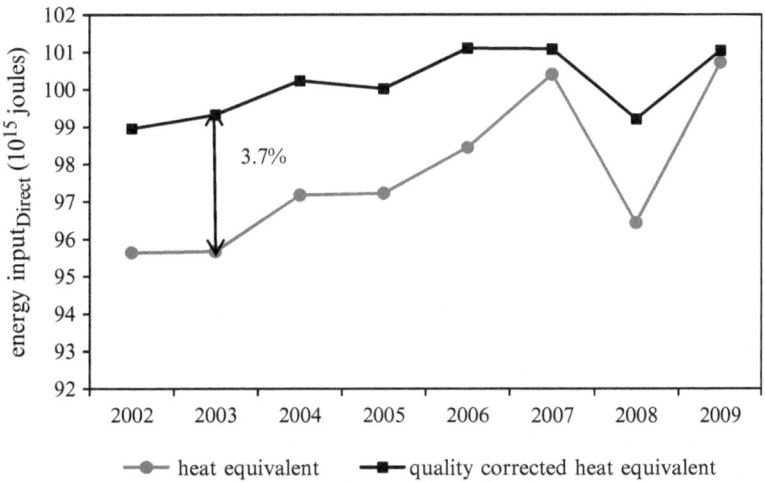

Fig. 4.9 Direct energy inputs for the Daqing oil field in heat equivalents and quality-corrected heat equivalents

where $E_{Indirect}$ refers to indirect energy inputs, M_{Total} and M_{Direct}, respectively, represent total money input and direct money input. E_{ins} is energy intensity for all of the industry in China and it is measured in tonnes of coal equivalents (ce) per 10^4 Yuan, which was 2.4 in the year of 2002. C_{CE-RC} is the conversion factor from ton of coal equivalent (tce) to the ton of raw coal (about 0.7143 kg tce/kg). C_{T-J} represents the conversion factor from ton of raw coal to joules. 2002 data is used as an example to illustrate how to get indirect energy input (Tables 4.7 and 4.8).

Table 4.8 Indirect energy input of the Daqing oil field in 2002

Indirect inputs	Unit	As energy
Indirect energy inputs	10^3 tonnes of coal equivalent	4,455
Indirect energy inputs	10^3 tonnes of raw coal	6,236
Indirect energy inputs (E_{Indirect})	10^{15} J	130.4

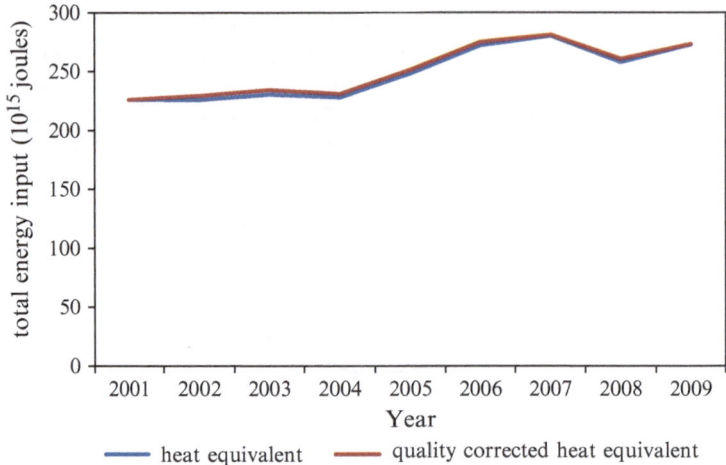

Fig. 4.10 Total energy inputs (heat equivalents and quality-corrected heat equivalents)

As Table 4.7 shows, we derived indirect monetary costs for the Daqing oil field from total monetary costs minus the direct monetary costs generated from physical units. Then, we got the indirect energy inputs by converting monetary units into energy units (Table 4.8).

3. Total Energy Inputs

   The second level of EROI analysis includes direct and also indirect energy inputs derived from financial data. Total inputs corrected for energy quality are shown in Fig. 4.10. There is only a very small difference between quality corrected and non-corrected total inputs because the quantity difference of direct energy inputs is so small (Fig. 4.11).

4.5.3   Results

We estimated that the standard energy return on investment ($EROI_{\text{stnd}}$) for the Daqing oil field decreased from about 10:1 in 2001 to 6.5:1 in 2009 (Table 4.9 and Fig. 4.12). The EROI derived in four different ways show the same decreasing trends, and EROI derived using heat equivalents is higher than when corrected through the Divisia index. However, EROI expressed as heat equivalents changes

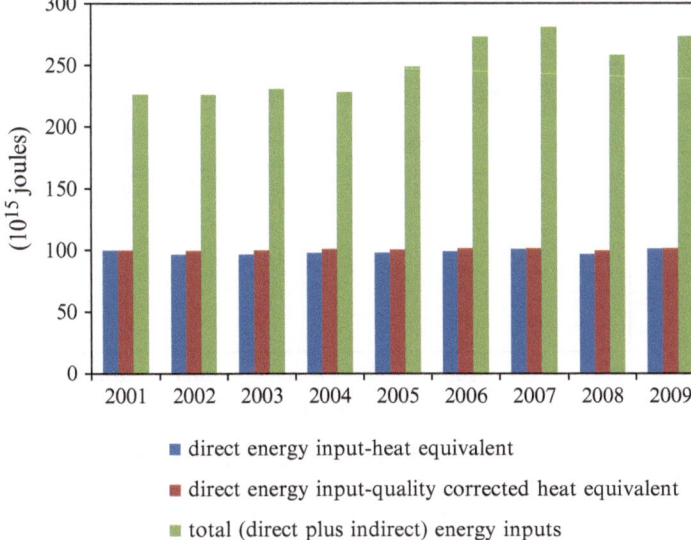

Fig. 4.11 Comparison of direct energy input (heat equivalents and quality-corrected heat equivalent) and total energy inputs (heat equivalents)

Table 4.9 EROI of the Daqing oil field

	2001	2002	2003	2004	2005	2006	2007	2008	2009
$EROI_{1,d}$	22.7	22.9	22.2	20.9	20.4	19.4	18.2	18.4	17.6
$EROI_{1,\,Qd}$	22.7	22.2	21.3	20.3	19.7	18.8	18.1	17.8	17.4
$EROI_{stnd}$	10.0	9.7	9.2	8.9	8.0	7.0	6.5	6.9	6.5
$EROI_{1,\,Qstnd}$	10.0	9.6	9.1	8.8	7.8	6.9	6.5	6.8	6.4

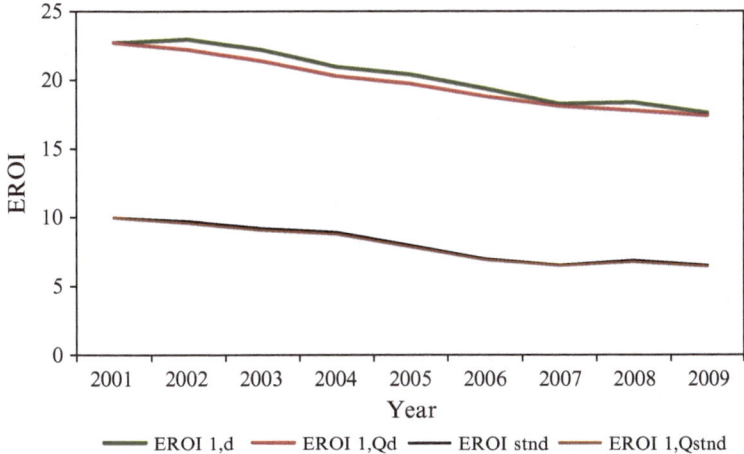

Fig. 4.12 The results of our assessment of the EROI of the Daqing oil field calculated in four different ways. The upper two lines do not include indirect energy costs and are less complete

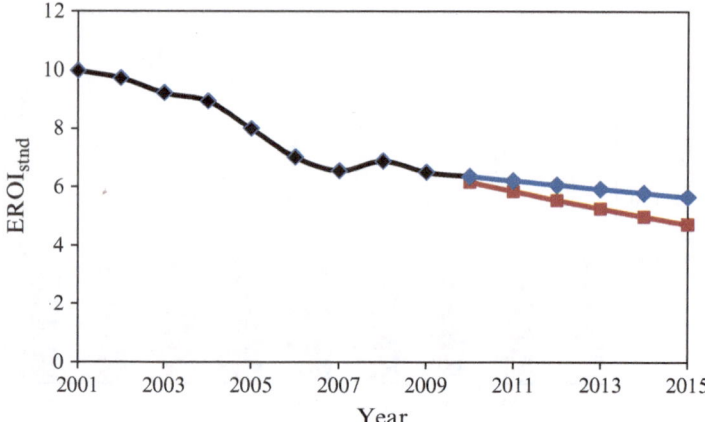

Fig. 4.13 History and forecast of the EROI$_{stnd}$ for the Daqing field. *Black line* is historical EROI$_{stnd}$; *Red line* is an extrapolation based on the best linear fit to trend; the *blue line* is an extrapolation of costs, assuming the Chinese government's goals for production are met, but energy costs continue to increase

less than when outputs and inputs are corrected for quality (Fig. 4.11). In addition, when the indirect costs are not included, the EROI appears to be higher, which, of course, is an artifact of the incomplete analysis.

Most often, an EROI analysis is determined by the data available. Since the Daqing oil field does not publish data on pollution we cannot include environmental data as an energy input. For example, the Daqing oil field increases the pressure of the polymers pumped into the ground each year which has large negative impacts on the environment (He 2008). If the negative externalities upon the environment were to be considered, the EROI of the Daqing oilfield would be decreased substantially compared to the value that we presented.

The declining trend of the EROI of the Daqing oil field demonstrates that oil and natural gas extraction is becoming more and more difficult even for very large and relatively well-managed oil fields such as the Daqing. The principal reason is that as oil field ages, it requires energy-intensive techniques, such as water and polymer injection under substantial pressure to recover the remaining oil. The productivity of any oil field eventually declines regardless of other circumstances. The reasons for the decline are varied, but the important thing is that it seems that depletion is a more powerful factor than technological improvements in Daqing. Another reason for the decline in EROI is that while the production of the Daqing has decreased slowly, the investment of funds and energy has increased almost linearly. We make a simple prediction in this book by extrapolating the output and input of Daqing oil field; our prediction is that the EROI is likely to continue declining over the next 5 years. We utilized the increasing rate of outputs and inputs to make a linear extrapolation, projecting the EROI$_{stnd}$ for the next 5 years (Fig. 4.13). If the decline in EROI continues to follow the present rate, it will reach very low values within one to two decades. In contrast, the outputs of the Daqing oil field are supposed to

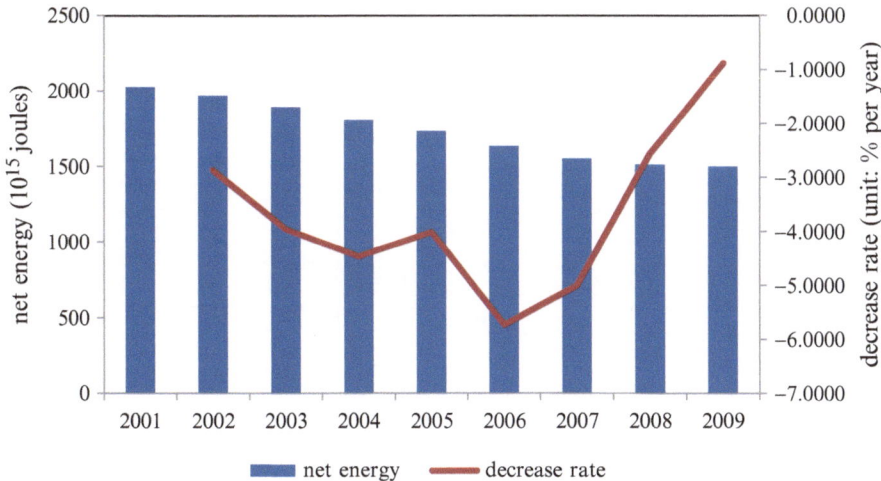

Fig. 4.14 Net energy and decrease rate of China's Daqing oil field

be determined by the national plan, which calls for the continued production of 296 million barrels per year. We accept this production for the moment, using it to make another extrapolation, but also assuming that inputs increased from 2001 to 2009. Under these assumptions, the EROI declines even if oil production remains flat. Since production of this oil field has been under the control of the government, which takes great pride in its ability to manage it, the decline in production is rather an embarrassment. The decline in EROI only makes matters worse, but is consistent with what is happening with nearly all other fossil fuels, as seen in the special volume of sustainability in which this analysis first appeared.

Net energy analysis related to EROI is of great importance, reflecting the amount of energy that can actually be delivered to society. We find that the net energy of the Daqing oil field has the same trend as EROI, both of which are declining, at 3.7% per year (Fig. 4.14). From the point of view of energy value, production will lose its significance if the net energy reaches zero, deeply impacting China's oil industry. Hence, both continuously decreasing EROI and net energy output indicate that the Daqing oil field is suffering from serious challenges now and will continue to do so in the future.

Chapter 5
Summary

The economic growth of China during the last several decades has been the envy of much of the developing world. It is usually assumed that this growth is due to the special effectiveness of Chinese workers or engineers or to special Chinese political actions. While these issues are certainly true a more important factor appears to be the effectiveness of China's industry in exploiting and harnessing energy. Take oil as an example, in the early of twentieth century, China began to use its first industrial well to extract oil with the production of one ton per day; from 1950 to 1952, along with increased exploration efficiency, crude oil production increased to 375 thousand tonnes; in 1998, the Chinese achieved the integrated operation of the industry including the upstream and downstream chains, with the production of 160 million tonnes, then China's oil production increased to 200 million tonnes in 2010.

5.1 The Importance of Energy to China's Economy

The economics of China has grown from almost nonexistent to existent, from small to large from weak to strong. Compared to the level of the Western world, however, China is just reaching the "mid-industrialization" stage. China's industrialization will not be completed until the 2030s (Chen et al. 2007) and much of China remains undeveloped and very poor. This means that China's GDP may keep growing rapidly until 2030. Some studies have predicted that China's future GDP growth will be 8% per year until 2020 (M.S. 2010), and 7% through 2030 (Li 2003; Li et al. 2005). The rapid growth of GDP in the future means that Chinese energy consumption will also continue to grow rapidly. Many studies have shown that there has been a close relation between Chinese economic development and its energy consumption over the last 30 years (Zhao and Fan 2007; Yuan et al. 2008; Li et al. 2011) (Fig. 5.1). This figure shows that China's GDP growth is clearly heavily dependent on the availability of massive energy resources. Both

L. Feng et al., *The Chinese Oil Industry: History and Future*,
SpringerBriefs in Energy, DOI 10.1007/978-1-4419-9410-3_5,
© Lianyong Feng, Yan Hu, Charles A.S. Hall, Jianliang Wang 2013

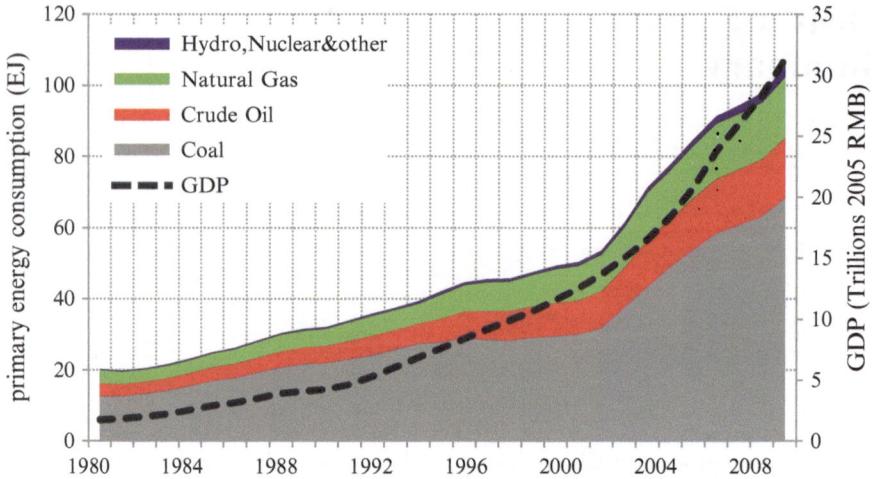

Fig. 5.1 The relation between the primary energy consumption and GDP of China. *Source*: China Energy Statistic Yearbook 2010 and China Statistic Yearbook 2010

time series of GDP and energy consumption, show the same increasing trend. Fossil fuels accounted for 93% (coal: 70%; oil: 19%; gas: 4%) of the primary energy consumption of China in 2010. The proportion of other energy sources, such as nuclear and renewable energy was only 7% as of 2010.

Li et al. (2005) and Kokichi et al. (2005) have forecasted that fossil fuels for China will continue being more than 90% of total energy use by 2020. The United Nations Development Programme (UNDP 2010) predicted that fossil fuels still would be the major energy, even when taking into account an Emission Control scenario aimed at reducing China's carbon emissions to the largest extent possible without affecting its economic growth. Therefore, China's future economic growth will remain heavily dependent on fossil fuels, unless they continue heavily investing in renewable energy. Thus, an important question is whether or not China will be able to find enough fossil fuels to meet its future energy demands or move to renewable energies, and if not China will be confronted with serious economic challenges in the future.

5.2   The Future for Fossil Fuel Supply to China

5.2.1   Oil

According to the latest national oil and natural gas resource assessment completed by the Chinese government in 2005, the ultimate recoverable reserves of Chinese oil resources are 21.2 billion tonnes (886 EJ).[1] The average URR estimated by several Chinese scholars, however, is only 13 billion tonnes (543EJ) (Chen 2003). Different estimates of China's oil URR influences the different predicted oil production using peak models, and also the different opinions of peak oil theory for China (See Chap. 3). Using an URR of 13.4 billion tons (560 EJ), Tang et al. (2010) predicted that Chinese oil production would reach its peak around 2015. Feng et al. (2008) used an URR of 15.6 billion tons (652 EJ) and forecasted that China's peak oil would occur in 2008 with a peak production of around 0.2 billion tons (8 EJ). However, Chinese oil demand is still rising rapidly. According to IEA's World Energy Outlook (IEA 2010), Chinese oil demand will increase from 0.4 billion tons (17 EJ) in 2010 to 0.7 billion tons (29 EJ) in 2030 even with the "New Policies" scenario.

Hence, importing oil has become increasingly important for China. In 2010, 53% of China's oil consumption was imported (IEA 2010). Chinese oil imports are likely to continue increasing in the future because China's oil production approach will reach its geological limits, and no longer increase significantly, even while China's consumption continues to grow. If these projections are correct, the gap between oil demand and domestic oil production would reach 0.6 billion tons (25 EJ) in 2030 (Wang and Feng 2011). Moreover, the world's oil export capacity may have entered its "peak plateau" during 2006–2016, with a peak export capacity of 2.8 billion tons (117 EJ). The decline of oil available to import and the probable increase in the import prices of oil will have serious impact on China's oil use. The Chinese government has noticed this problem and has started taking measures to cope with the country's increasing oil gap, through the development of a "Coal-To-Liquids" processes (CTL) (a process by which liquids such as gasoline and diesel are made from coal), mainly because the government believes that China is rich in coal. In 2009, China's CTL capacity reached 1.6 million tons per year (0.07 EJ/year) (Yang 2010a, b). China's CTL capacity is forecasted to reach 12 million tons per year (0.5 EJ/year) by 2015, 50 million tons per year (2 EJ/year) by 2020, and to keep expanding further afterwards (Yang 2010a, b). Unfortunately, replacing oil with coal is not a workable solution in the long term because coal resources are also non-renewable. Conversion ratios for CTL are generally estimated to be between 1 and 2 barrels of liquid/ton of coal (Höök and Aleklett 2010).

5.2.2   Natural Gas

China has even less natural gas resources than oil. According to estimates by Chinese scholars, China's natural gas URR amounts to 6–13 tcm (233–506 EJ), with an average URR equal to 9.6 tcm (373 EJ) (Zhang 2002b), far less than the 22

---

[1] We convert all physical units (ton, bcm, etc.) into joules using conversion factors (1 kg crude oil= 41.8 million joules; 1cubic meter=38.9 million joules; 1 kg raw coal= 20.9 million joules).

tcm (856 EJ) estimated by the Chinese government (Dai et al. 2008). Even if using the official government URR, production in 2020 will be only 165 bcm (6 EJ), and will reach its peak in 2043 with a peak production of 218 bcm (8 EJ). Meanwhile, according to the 12th Five-year Plan of the Chinese government, projected natural gas demand will be much higher reaching 260 bcm (10 EJ) by 2015 and 350–400 bcm (14–16 EJ) by 2020. Therefore, the gap between demand and supply in 2020 will reach 185–235 bcm (7–9 EJ). If the Chinese Scholars' low URR estimate is correct, the rate of extraction in 2020 would exhaust reserves in 15 years.

In 2006, China became an importer of natural gas for the first time, in the form of liquefied natural gas. It became a net importer in 2007 and by 2009, the net imports accounted for 5% of the total consumption. In 2009, China began importing natural gas through a pipeline from Turkmenistan on the very last day of the year (Hydro-carbon Asia 2010). China's gas imports have been increasing rapidly since 2006, reaching 12.3 bcm (0.5 EJ) in 2010 (BP 2011). Based on current natural gas import plans, China can import only114 bcm (4 EJ) of natural gas by 2020 (including LNGs and pipeline gas) (Qiu and Fang 2009). As a consequence, China will need to find 71–121 bcm (3–5 EJ) of natural gas to meet the domestic demand in 2020. Hence, it would be very hard to find enough natural gas resources to meet the Chinese demand, especially if it continues to grow.

5.2.3   Coal

According to the latest survey on national coal resources, completed in 1997, China's total coal resources are 5.6 trillion tons [117×10^{21}(ZJ)]. However, the recoverable reserves are thought to be only 189 billion tons (4 ZJ). Based on this official data, Lin and Liu (2010) predicted that China's coal production would peak in 2025, and decline thereafter, with a peak production of 3.8 billion tons (0.08 ZJ). An increasing number of scholars and institutions analyzing the official coal data believe that the recoverable reserves are overestimated. Rutledge (2010) believes that the recoverable coal reserves are 139 billion tons (2.9 ZJ). Mohr and Evans (2009) claim that the recoverable coal reserves are actually 130–146 billion tons (2.7–3.1 ZJ) and that production will peak in the period of 2010–2017. The Energy Watch Group (2007) concluded that the recoverable coal reserves are only 120 billion tons (2.5 ZJ), with a production peak in 2015, followed by decline. Hence, we believe that Lin and Liu's estimated date for China's coal production reaching a peak in 2025 is most likely to be too optimistic, and that the decline will actually begin years earlier.

Compared with exaggerated coal resources and overestimated coal production, China's coal demand is generally underestimated. For example, a few years ago, most forecasts for coal demand in 2010 were between 2.5 billion tons (52 EJ) and 2.9 billion tons (61 EJ) (Lin et al. 2007; Wang and Li 2008; Yu and Deng 2008), including an estimate of 2.6 billion tons (54 EJ) from the 11th (from 2006 to 2010) Five-Year Plan. However, actual coal consumption in 2010 was 3.2 billion tons

(71 EJ), which is 32% higher than the official plan (BP 2012). Shealy and Dorian (2010) also claim that most forecasts of China's future coal demand are too low. Using a relatively conservative annual GDP growth target of 6.5% for the next 15 years, they forecasted that coal demand will reach 4.1 billion tons (86 EJ) in 2015, 9% higher than official data. Their projected coal demand in 2025 is 35.5%, 36.3% and 47.6% higher than the results of the International Energy Agency (IEA) (2010), EIA (2010) and the NDRC (2009), respectively.

China has been a net coal importer since 2009, and it seems likely that China's coal imports will continue to increase in the future. According to our previous analysis, even with optimistic estimates, China's coal production will peak by 2025. However, even with conservative assumptions about Chinese GDP growth, China's coal demand will keep increasing rapidly in the future. If we compare a forecasted demand of 6.1 billion tons (127 EJ) by 2025 with an estimated supply of 3.8 billion tons (79 EJ), China's net coal imports will reach 2.3 billion tons (48 EJ) by 2025, which means a 37% degree of dependence on foreign coal. According to EIA statistics, total coal trade in international markets was 0.9 billion tons (19 EJ) in 2009. Therefore, in order to meet China's coal demand, even assuming that other coal importers don't import any coal from the international market, world coal trade must increase by 144% over the next 15 years. After 2025, with the high probability of rapidly declining Chinese domestic coal production, the amount of world coal trade must increase even more rapidly than now. It seems very hard to imagine that world coal trade will increase rapidly enough to meet Chinese coal import demand by then if the economy continues to grow rapidly.

5.3 Restrictions of Energy on Chinese Economic Growth in the Future

5.3.1 Peak Fossil Fuels

China's economic growth in the future will probably remain heavily dependent on increasing consumption of fossil fuels which have to depend on an adequate supply of inexpensive fossil fuels. At this point, the future of China's internal production of fossil fuels is uncertain. We are concerned about the possibility of peak fossil fuels and have estimated the possible trends of fossil fuels supplies using the multi-cyclic Generalized Weng model (Chap. 3) with low and high URR scenarios. The results for both low and high scenarios are almost the same. According to our analysis, the peak production of fossil fuels will be 3.4 billion tce by 2020 with low URR estimation (Fig. 5.2), and the peak production of fossil fuels will be 3.4 billion tce by 2021 with high URR estimations (Fig. 5.3).

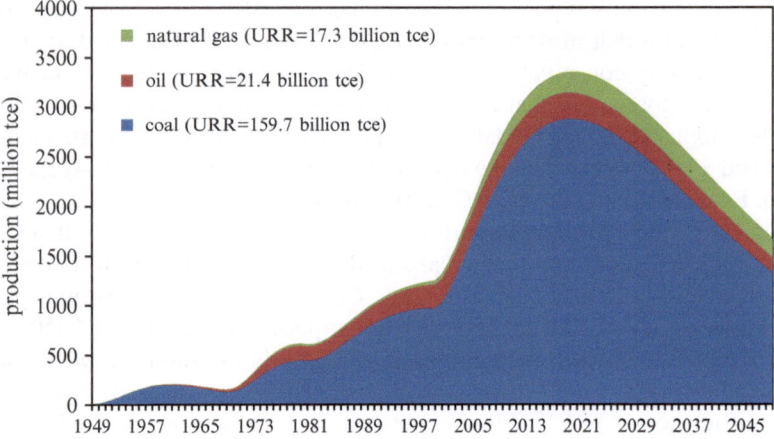

Fig. 5.2 Probable fossil fuel supplies for China with low URR scenario in the future. Low URR scenario refers that (1) URR for natural gas is 17.3 billion ton coal equivalent (tce), (2) URR for oil is 21.4 billion tce, (3) URR for coal is 159.7 billion tce

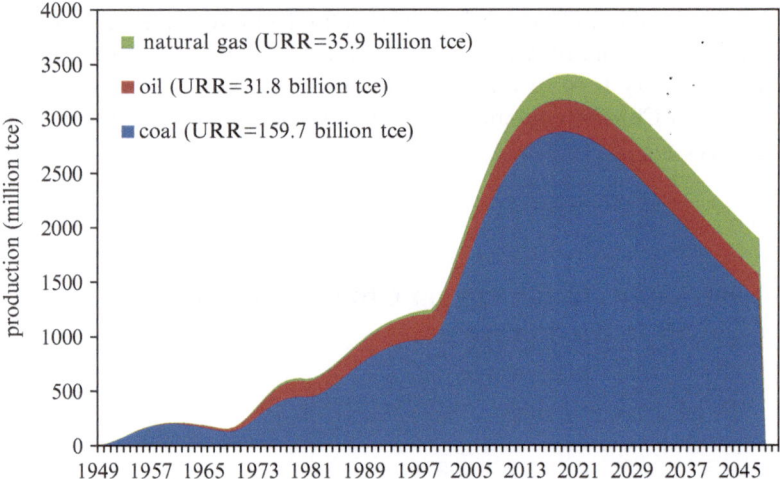

Fig. 5.3 Probable fossil fuel supplies for China with high URR scenario in the future. High URR scenario refers that (1) URR for natural gas is 35.9 billion ton coal equivalent (tce), (2) URR for oil is 31.8 billion tce, (3) URR for coal is 159.7 billion tce

5.3.2 *Declining EROI*

EROI analysis is useful in at least two ways: first for getting energy itself, e.g., the normal process of oil exploration, development and production from an oil field or province, and second, more generally, for the energy required to maintain and develop an economy or a society. However, over most of the last 30 years, EROI has not received the attention that we think it should have, probably because energy prices started a long decline from 1980 until 2005. With the large increase in the price of energy, most researchers and also public officials have become considerably interested in EROI and models of peak production. Additionally, EROI is bound up with the "net" energy approach, describing numerically how much energy is left to power our industrial society and modern civilization after extracting, processing and delivering the energy.

Over time, with the development of technology, Chinese society has consumed and invested an increasing amount of energy to find and produce energy, i.e., drilling more footage in the fields tends to lead to higher demand for energy and further depletion, not to higher rate of production. The $EROI_{stnd}$ estimates for China's oil and gas extraction are: (1) China's oil and natural gas extraction showed the slow declining trend from the maximum value at 14:1 in 1996 to 10:1 in 2010 with the decreasing rate of 2.6%; (2) $EROI_{stnd}$ for Daqing oil field declined continually from 10:1 in 2001 to 7:1 in 2009. From the point of view of energy quality, which refers to the ability of one joule to generate economic output, we know that the quality of China's fossil fuels has been going down. Additionally, EROI analysis provides a useful method to compare the output of different energy types with the same amount of energy inputs.

5.3.3 Energy Efficiency

Enhancing energy efficiency has become China's important focus in order to face the energy challenges of the future. Improving energy efficiency is going to play an extremely important role in mitigating the contradiction of supply and demand and economic growth. Energy intensity is a measure of the energy efficiency of a nation's economy. It is calculated as units of energy per unit of GDP. Peaked fossil fuels will not be a problem if the Chinese energy consumption per GDP declines substantially. That is, if GDP increases with the same or less energy consumption because energy intensity declines in the same period.

Since China's implementation of reform and opening up policy in 1978, the change rate of China's economy has been higher than the change rate of primary energy consumption except for 2003 and 2004 (Fig. 5.4) leading to the decreasing of energy intensity.

Reviewing historical data of China's energy intensity, we found that (1) the energy intensity dropped slowly from 1988 to 1991 caused by the decline of industry value added during this period; (2) the energy intensity increased from 2002 to 2005 because of the development of heavy industry, the larger scale of infrastructure construction, and the consumption of housing and automobiles (Fig. 5.5).

The general declining trend of energy intensity caused debate about data authenticity. However, we found that the change rate of energy intensity fluctuated between −0.02 and −0.08 (except for 2003 and 2004) with a clear periodicity. Therefore, we do not think that China's economic data has problems because of this general declining trend (Fig. 5.6).

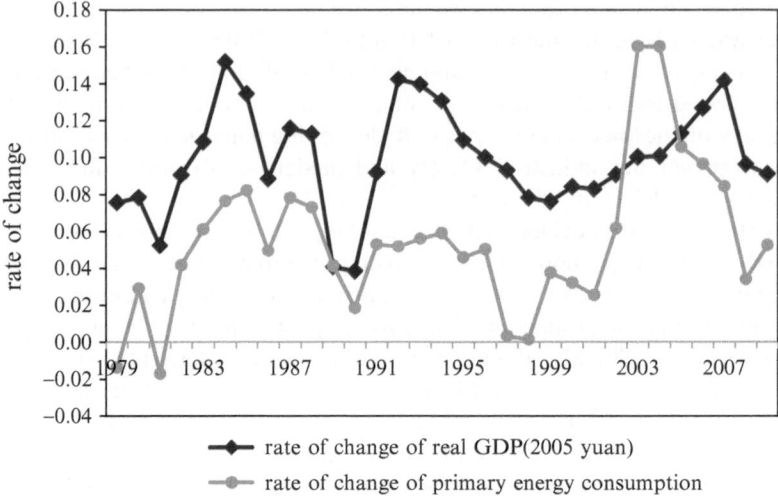

Fig. 5.4 Rate of change of real GDP and primary energy consumption of China. *Source*: China Energy Statistic Yearbook 2010 and China Statistic Yearbook 2010

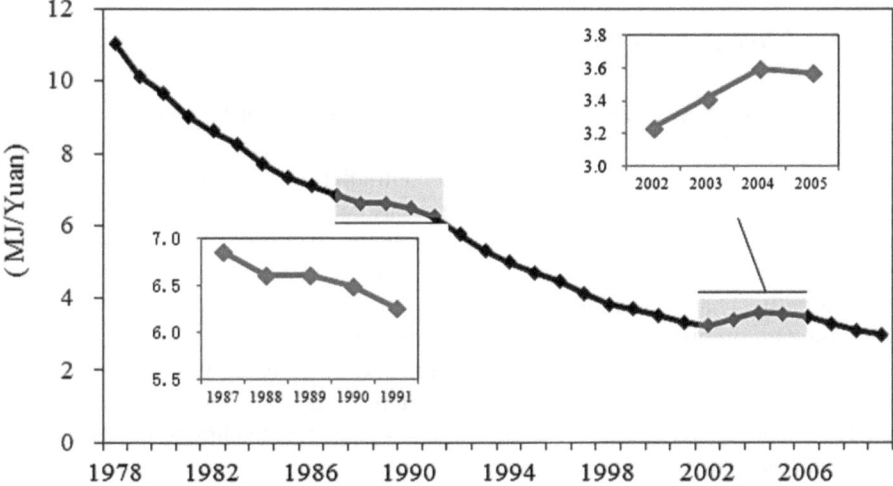

Fig. 5.5 Energy intensity in China (primary energy consumption/real GDP (2005 Yuan)). *Source*: China Energy Statistic Yearbook 2010 and China Statistic Yearbook 2010

We are not too optimistic about future high economic growth of China from the perspective of supply challenges, peak fossil fuels, and EROI analysis. However, we are optimistic about the recent energy efficiency tendency of China. We do not have to be worried about economic growth for the future if the rate of depletion of fossil fuels is less than the increased rate of energy efficiency because less energy consumption could be able to create more GDP to promote economic growth as

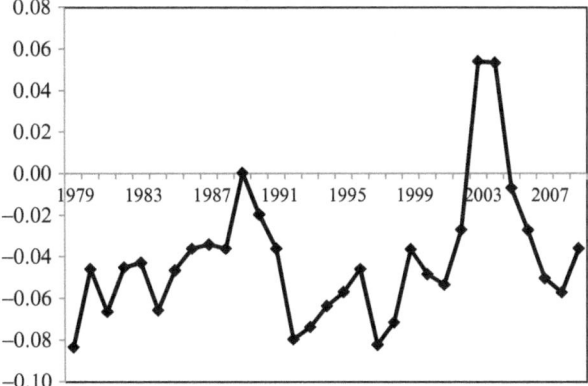

Fig. 5.6 Rate of change of energy intensity change rate for China. *Source*: China Energy Statistic Yearbook 2010 and China Statistic Yearbook 2010

long as it is used efficiently. One analysis of China's energy efficiency (Zhang and Wang, 2008) found that the energy intensity has an inverse relation with coal consumption, that is, the larger the share of coal in primary energy consumption, the lower the energy efficiency. Nevertheless, the dominance of the coal industry in China will not change in the short term just by EROI analyses like ours, although we are hopeful that China will continue moving towards renewable energies. Lastly, China's heavy high energy-intensive industry has entered the development stage with rapid speed, which causes growth in added value of industry and energy consumption. Consequently, energy efficiency and indeed the entire future of the Chinese economy are full of uncertainty. The country's intellectual and political leaders would be well advised to consider the possibility that the past pattern of very high economic growth might not continue indefinitely.

References

A R.H., 2010. "Historical Development of Jilin Oilfield" *Overview of Songyuan Cuture,* 9 (in Chinese).available at: http://www.sywhw.org.cn/cjsywhc/sywh/informations/20100908/20165.shtml

ASPO-China, 2005. "The Group for the Global Energy System-China". www.cup.edu.cn/peakoil. (in Chinese)

Bai, Z.Y., 2009. "Petroleum Memory", University of Petroleum Publishing House, BeiJing. (in Chinese)

Berndt, E.R., 1978. "Aggregate Energy, Efficiency and Productivity Measurement" *Annual Review Energy,* 3, 225–273

Berndt, E.R., 1990. "Energy Use, Technical Progress and Productivity Growth: A Survey of Economic Issues" *J. Prod. Adnal.,* 2, 67–83

BP. 2010.BP Statistical Review of World Energy. http://www.bp.com/statisticalreview

BP, 2011.BP Statistical Review of World Energy. http://www.bp.com/statisticalreview

BP, 2012. BP Statistical Review of World Energy. http://www.bp.com/statisticalreview

Brown, M.T., Ulgiati, S., 2004. "Energy Quality, Emergy, and Transformity: H.T. Odum's Contributions to Quantifying and Understanding Systems" *Ecol. Model.,* 178, 201–213

Bullard C.W., Herendeen, R.A., 1975. "The Energy Cost of Goods and Services", *Energy Policy,* 12, 268–278

Chang, L.T., 2005. "Production of Bohai oilfield exceeds ten million cubic meters" *Oil Field Equipment,* (2):78. (in Chinese)

Chen, Y.Q., Hu, J.G., 1995. "The Weibull Model of Predicting Production and Recoverable Reserves of Oil and Gas Field" *Xinjiang Petroleum Geology,* 16(3), 250–255. (in Chinese)

Chen, Y.Q., 1996. "Derivation and Application of Generalized Weng's Model" *Natural Gas Industry,* 16(2), 22–26. (in Chinese)

Chen, Y.Q., Yuan, Z.X., 1997. "The Foundation and Application of Long-Normal-Distribution Predicting Model" *Acta Petrolei Sinica,* 18(2), 84–88. (in Chinese)

Chen, X.S., Wang, J.C., 1999. "The Effects of China's Crude Oil and Oil Products Price Reform Policy" *International Petroleum Economics,* (1), 19–24.(in Chinese)

Chen, Y. Q., 2003. "Forecast of Oil Production and Recoverable Reserve in China" Oil Forum, 1, 26–31(in Chinese)

Chen, J.G. et al., 2007. "China Enters the Last Half Phase of Mid- Industrialization Stage" *Newspaper of Institute of Economics Chinese Academy of Social Sciences.* 9/27

Chen, G., 2008. "Review and Thinking of Oil Industry Reform and Opening 30 Years" (in Chinese) http://energy.people.com.cn/GB/8356135.html

China Enterprise Management Annual, Ltd, SLOF(Shengli Oil Field), Sinopec., 2005. (in Chinese)

China Energy Statistical Yearbook, 1995-2009. Energy Statistics Department of National Bureau of Statistics of China. China Statistics Press: Beijing, China

Chu, F.J., An, Y.P., 2004. "A Difficult Business to Cover the History of Earthshaking Million People - Review of the 55th Anniversary of the Oil Industry" *China petrochem*, 10. (in Chinese)

Cleveland, C.J., Costanza, R., Hall, C.A.S., Kaufmann, R., 1984. "Energy and the United States Economy–a Biophysical Perspective" *Science*, 225, 890–897

CNOOC. 2009. CNOOC Group Annual Report 2009. (in Chinese) http://www.cnooc.com.cn/data/upload2/xiazai/2009ARe.pdf

CNPC. 2009.CNPC Group Annual Report 2009. (in Chinese) http://www.cnpc.com.cn/resource/cn/other/pdf/2009%e9%9b%86%e5%9b%a2%e5%b9%b4%e6%8a%a5%e7%bb%88%e7%a8%bf.pdf

CNPC Website. 2010. CNPC Group Annual Report 2009. (in Chinese) http://www.cnpc.com.cn/resource/cn/other/pdf/2009%e9%9b%86%e5%9b%a2%e5%b9%b4%e6%8a%a5%e7%bb%88%e7%a8%bf.pdf

Cui, Y.K., 2000. "Vaule Chain Analysis and Design Of Liaohe Oilfield" *Dalian University of Technology*, (in Chinese)

Cui, H.W., 2004. "Progress and Development of Zhongyuan Oilfield Yearbook" *Yearbook Information and Research*, 06, 18–20. (in Chinese)

Dai, J. X., Ni, Y. Y., Zhou, Q. H., Yang, C., Hu. A. P., 2008. Significances of Studies on Natural Gas Geology and Geochemistry for Natural Gas Industry in China. *Petroleum Exploration and Development*, 35(5), 513–525

Daniel, Y., 1992. "The Prize: the Epic Quest for Oil, Money and Power", Shanghai Translation Publishing House. (in Chinese)

Deng, S.H., 1992. "Brief Introduction of Modern Oil Industry in Xinjiang" *Journal of Xinjiang University (Philosophy and Social Science Edition)*, 4, 52–58. (in Chinese)

Energy Watch Group. 2007. Coal: Resources and Future Production. EWG-Series No 1/2007, March. Available at: http://www.energywatchgroup.org/fileadmin/global/pdf/EWG_Report_Coal_10-07-2007ms.pdf

Energy Information Agency (EIA). 2010. "International Energy Outlook". Available at: http://www.eia.gov/oiaf/ieo/index.html

Energy Information Administration (EIA), 2011.http://www.eia.gov/countries/cab.cfm?fips=CH (accessed on 25 November 2011)

Feng, L. Y., Li, J. C., Pang, X. Q., 2008. "China's Oil Reserve Forecast and Analysis Based on Peak Oil Models" Energy Policy, 36(11), 4149–4153

Feng, L.Y., Hu, Y., Sun, W.M., 2009. "Several Proposals on Present Pricing mechanism of refined oil product in China" *Price:Theory&Practice*, (7), 41–42. (in Chinese)

Feng, LY., Wang, J.L., Zhao, L., 2010. "Construction and Application of A Multi-Cycle Model in the Prediction of Natural Gas Production" *Natural Gas Industry*, 30(7), 110–112. (in Chinese)

Fu, X., 2010. "International Experience and Enlightenment of Strategy Oil Reserves Management" *Macroeconomic Management*, 2, 45–47. (in Chinese)

Gagnon, N., Hall, C.A.S., Brinker, L., 2009. "A Preliminary Investigation of Energy Return on Energy Investment for Global Oil and Gas Production. Energies" *Energies*, 2, 490–503

Guan, Q.Y., 2010. "Peak Oil will Change International Politics and Economic Domain" *China Securities Journal*, 6.21.,page A04. (in Chinese)

Guilford, M.C., Hall, C.A.S., Connor, P.O., Cleveland, C.J., 2012. "A New Long Term Assessment of Energy Return on Investment (EROI) for U.S. Oil and Gas Discovery and Production" *Sustainability*, 3, 1866–1887

Hall, C.A.S., Cleveland, C., Berger, M., 1981. "Yield per effort as a function of time and effort for United States petroleum, uranium, and coal" Energy and Ecological Modelling; Mitsch, W.J., Bosserman, R.W., Klopatek, J.M., Eds.; Elsevier Scientific: Amsterdam, The Netherlands

Hall, C.A.S., Day, J.W., 2009. "Revisiting the Limits to Growth After Peak Oil", American Scientist, 97, 230–237

Hallock, J.L., Tharakan, P.J., Hall, C.A.S., Jefferson, M., Wu, W., 2004. "Forecasting the limits to the availability and diversity of global conventional oil supply", Energy, 30, 2017–2018

Hamilton, J., 2012. "Maugeri on Peak Oil", Energy Bulletin, Available on line: http://www.energybulletin.net/stories/2012-07-19/maugeri-peak-oil

Han, D.K, 2010. "Status and Challenges for Oil and Gas Field Development in China and Directions for the Development of Corresponding Technologies" *Engineering Sciences,*12 (5), 51–57. (in Chinese)

Hao, H.Y., Hu, Y., Feng, L.Y., Zhao, L., 2008. "The Truth and Lies of Peak Oil-Attitudes and Behaviors of International Oil Companies" *Social Sciences Abroad,* 5, 78–83. (in Chinese)

He, L., 2008. "Current Situation and Prospect for Oil Recovery Techniques after Polymer Flooding in Daqing Oilfield", *Oil Drilling & Production Technology*, 30(3), 1–6

Hu, J.G., Chen, Y.Q., 1995a. "A New Model of Predicting Production and Recoverable Reserves of Oil and Gas Field" *Xinjiang Petroleum Geology, Acta Petrolei Sinica,*16(1),79–86. (in Chinese)

Hu, J.G., Chen, Y.Q., 1995b. "Derivation, Application and Discussion of t-Model" *Natural Gas Industry,* 15(4), 26–29. (in Chinese)

Hu, J.G., Chen, Y.Q., 1997. "The Foundation and Application of Hu-Chen Predicting Model" *Natural Gas Industry*, 17(5), 31–34. (in Chinese)

Hu, J., Wu, W.J.,2007. "The Conflict between Mining Rights of Oil and Gas Resources and Land Property Rights: a Case Study on the Development of Oil and Gas Resources in North Shaanxi" *Resources Science*, 29(03), 8–16. (in Chinese)

Hu, S.L., 2010. "The Shadow of Peak Oil" *China Energy Newspaper,* 9/27: 013. (in Chinese)

Hu, W.R., 2011. "Treat Peak Oil by Views of Development" *China Petrochem,* 9, 32. (in Chinese)

Hu, X.J., Wang, L.N., 2011. "The Discovery of the Karamay Oilfield: An Interview with Tian Zaiyi" *The Chinese Journal for the History of Science and Technology*, 2, 267–274. (in Chinese)

Huang, F.S., Zhao, Y.S., Liu, Q.N., 1987. "A New Model of Dynamic Prediction of Oil Field" *Petroleum Geology & Oilfield Development in Daqing*, 6(4), 55–62. (in Chinese)

Huang, Y.C., Chen, Z.B., 2007. "Oil Geopolitics in High Price and China s Oil Trade Patterns" *Resources Science*, 29 (1), 172–177. (in Chinese)

Huang, X.F., 2011. "The Dynamic Assessment of National Oil and Gas Resources" *Economic Daily News*, 11/25. (in Chinese)

Hubbert, M.K. 1969. Energy Resources. In, P. Cloud (Ed.), Resources and Man. Freeman, San Francisco, p. 157–242

Höök, M., Aleklett, K., 2010. "A Review on Coal To Liquid Fuels and Its Coal Consumption" *Int. J. of Energy Research*, 34(10), 848–864

Hydrocarbon Asia. 2010. "Natural Gas Market in China: On the Fast Track to Becoming a Giant International Player?". Available at: http://www.hcasia.safan.com/mag/hca1210/r26.pdf

International Energy Agency (IEA), 2010. "World Energy Outlook". Available at: http://www.worldenergyoutlook.org/

Jiang, B., Zeng, F.G.,fanguang, She, S.C., 2008. "China Guangdong Provincial Party Committee Propaganda Department of the Task Force, A New Step in the Emancipation of the Mind" Qiu Shi, 07. (in Chinese)

Jiang, S.C., Shi, J., 2009. "Laojunmiao Development of Oil Yield Four Decades Creating a Miracle" *China Petroleum Daily*, 11, 06. (in Chinese)

Journal editor, 2003. "The Report for China's Oil and Gas Resources Strategy to Sustainability", *National Land& Resources Information*, 2, 37–41.(in Chinese)

Journal editor, 2008. "The Result of new national oil and gas resource assessment" *Henan Land &Resources*, 9, 42 (in Chinese)

Kang, S.L., 2001. "Endless Fate of Oil" *Underground Flashes,* 2, 16–28. (in Chinese)

Kang yuzhu. 2003. Geological characteristics of the formation of the large Tahe oilfield in the Tarim basin and its prospects. Chinese Geology, (03):315–319. (in Chinese)

Kaufmann, R.K., 1994. "The Relation Between Marginal Product and Price in U.S. Energy Markets: Implications for Climate Change Policy" *Energy Economic*, 16, 145–158

King, C. W., Hall, C.A.S., 2011. "Relating Financial and Energy Return on Investment" *Sustainability*, 3, 1810–1832

Kokichi, I., Li, Z. D., Komiyama, R., 2005. "Asian Energy Outlook to 2020: Trends, Patterns and Imperatives of Regional Cooperation. Research and Information System for Developing Countries (RIS)". Available at: http://www.newasiaforum.org/dp93_pap.pdf

Li, Z. D., 2003. "An Econometric Study on China's Economy, Energy and Environment to the Year 2030" *Energy Policy*, 31(11), 1137–1150

Li, J. M., Liu, S. Z., Li, D.X., Ma, S.P., 2004. "Exploratory Situation and Development Tendency of China's Natural Gas" *Natural Gas Industry*, 24(12), 1~4(in Chinese)

Li, Z. D., Kokichi, I., Komiyama, R., 2005. "Energy Demand and Supply Outlook in China for 2030 and a Northeast Asian Energy Community" . Available at: http://eneken.ieej.or.jp/en/data/pdf/300.pdf

Li, N.F., 2008. "The Thinking of China's Energy Strategies based on Peak Oil" *Journal of BPMTI*, 2, 4–7. (in Chinese)

Li, G.Y., 2009. "Review and Prospect of China's Oil and Gas Exploration in the Last 60 Years" *Oil Forum*, 5, 1–8. (in Chinese)

Li, Z.X., 2010. "The Memory of Yumen Oilfield" *Oil Political Studies*, 3, 70. (in Chinese)

Li, F., Dong, S. C., Li, X., Liang, Q. X., Yang, W. Z., 2011. . "Energy Consumption-Economic Growth Relationship and Carbon Dioxide Emissions in China" *Energy Policy*, 39 (2), 568–574

Lin, B.Q., 2007. "Bad and Good News Of Peak Oil" *21 shi ji jing ji bao dao*, 7/30, 031. (in Chinese)

Lin, B. Q., Wei, W. X., Li, P. D., 2007. "China's Long-Run Coal Demand: Impacts and Policy Choice". *Econ. Research J.*, 42(2), 48–58 (in Chinese)

Lin, B. Q., Liu, J. H., 2010. "Estimating Coal Production Peak and Trends Of Coal Imports in China" *Energy Policy*, 38(1), 512–519

Liu, J.P., Ma, Y., 2006. "On the Whole Establishment of Yan-Chang Oil Deposit in the Late Qing Dynasty" *Journal of Baoji University of Arts and Sciences(Social Science Edition)*,6, 62–64. (in Chinese)

Luo ming. 2004. Development Strategy of CNOOC and Quoted Companies. Economic Analysis of China Petroleum and Chemical Industry, (11):28–34. (in Chinese)

Lv, Z.Y., 1983. "The Issues Illustrated By U.S. and British Imperialism Exclusive China "Kerosene" Market" *Collected Papers of Historical Science*, 04. (in Chinese)

Maugeri, Leonardo June 2012. Oil: The Next Revolution. The Unprecedented Upsurge of Oil Production Capacity and What It Means for the World. Discussion Paper 2012-10, Belfer Center for Science and International Affairs, Harvard Kennedy School

Meadows, D.H., Meadows, D.L., Randers, J., Behrens, W.W., 1972. Limits to Growth, New York: New American Library

Ming, G.Z., Yunm S.Q., 2006. "The History of Oil" Petroleum Industry Press, Beijing. (in Chinese)

Ministry of Land and Resources (MLR), 2005. National Development and Reform Commission (NDRC), Ministry of Finance(MF). Assessment of Oil Resources (unpublished). (in Chinese)

Ministry of Land and Resources of the People's Republic of China (MLRPRC). 2010. "Communique On Land and Resources of China 2009" *National land & Resources Information*, 7, 4–12. (in Chinese)

Mohr, S. H., Evans, G. M., 2009. "Forecasting Coal Production until 2100" *Fuel,* 88(11), 2059–2067

Morgan Stanley Research Asia/Pacific (M.S.), 2010. "Chinese Economy through 2020: Not Whether but How Growth Will Decelerate" *China Economics*, 9/20

Murphy, D.J., Hall, C.A.S., Dale, M., Cleveland, C.J., 2011. "Order from Chaos: A Preliminary Protocol for Determining the EROI of Fuels" *Sustainability*, 3, 1888–1907

National Development and Reform Commission (NDRC), 2009. "2050 China Energy and CO2 Emissions Report" Published by Science Press in Beijing, China

National Development and Reform Commission (NDRC), 2011.http://www.sdpc.gov.cn/zfdj/default.htm (accessed on 25 November 2011)

NBSC. 2009. Statistical Communiqué of the People's Republic of China on the 2009 National Economic And Social Development(in Chinese). Available at: http://www.stats.gov.cn/tjgb/.

Niu, Z.Z., 2008. "Conceptions on Some Problems of the Renewable Energy LawOn Some Problems of the Renewable Energy Law" *Energy Conservation & Environmental Protection*, (9), 14–16. (in Chinese)

Oil Price Trends, 2007.*International Petroleum Economic*, 7, 73

Oil Price and Related Index Trends, 2008.*International Petroleum Economic*, 4, 92

Oil Price and Related Index Trends, 2009.*International Petroleum Economic*, 7, 87

Pu, M., 2010. "Developments in China's Oil and Gas Pipelines 2009" *International Petroleum Economics*, 18(3), 14–20. (in Chinese)

Pu, M., Ma J.G., 2011. "Developments in China's Oil and Gas Pipelines 2010" *International Petroleum Economics*, 19(3), 26–34. (in Chinese)

Qian, J., 2004. "Some Issues on China's Petroleum Resource Potential" *Oil & Gas Geology*, 25(4), 363–369. (in Chinese)

Qian, B.Z., 2007. "World Oil Production will be Peak: Right Or Not?" *China Petrochemical Industry*, 3, 52–54. (in Chinese)

Qian, B.Z., 2008. "China will Face Peak Oil in 2015" *Natural Gas and Oil*, 26(2), 3. (in Chinese)

Qian, B.Z., 2009. "The Research of China's Oil and Gas Peak Production" *Petroleum Refinery Engineering*, 39, 6. (in Chinese)

Qian, B.Z., 2010. "A New Competitive Pattern is presented in Refining Industry of China (The 1st Part)" *China Petroleum*, 09, 01. (in Chinese)

Qin, J., 2007. "Strategy Research on Internationalization" *Chian Petroleum Enterprise*, (8), 62–64. (in Chinese)

Qiu, Z. J., Fang, H., 2009. "Surging of Natural Gas in China: A New Journey of China's Petroleum Industry" *Natural Gas Industry*, 29(10), 1–4 (in Chinese)

Qu, G.H., 2011. "A Study into China's NG Consumption during Twelfth Five-Year Plan Period, Main Characteristics and Development Direction" *Sino-Global Energy*, 16(2), 1–7. (in Chinese)

Rutledge, D., 2010. "Estimating Long-Term World Coal Production with Logit and Probit Transforms" *Int. J. Coal Geol.*, 85(1), 23–33

Sinopec.2009. Sinopec Group Annual Report 2009. (in Chinese) http://www.sinopecgroup.com/gsjs/Doc/GroupAnnualReport2009.pdf

Shealy, M., Dorian, J. P., 2010. "Growing Chinese Coal Use: Dramatic Resource and Environmental Implications" *Energy Policy*, 38(5), 2116–2122

Shi, E.K., 1995, "Shien Kang' Description for China's Oil Industry" Petroleum Industry Press, Beijing. (in Chinese)

Song, L.S., 2005. "The Whole Story of Industry Learning from Daqing" People's Publishing House, HuBei. (in Chinese)

Song, W.J., Jiang, T.W., 2008. "Status of Petroleum Exploration and Production in the Tarim Basin and Guarantee of Supply for the West-to-East Gas Pipeline Project" *Natural Gas Industry*, 28(10), 1–4. (in Chinese)

Song, X.J., 2010. "Capability and Production of Ethylene of China in 2009 have Reached Ten Millions Tonnes" *Qilu Petrochemical Technology*, 01. (in Chinese)

Su, J., Luo, N., Cheng, W.D., 2010. "An Shuxing totalistic Thinking of Development of Huabei Oilfield Company in the Future" *International Petroleum Economics*, 05, 86–90. (in Chinese)

Sun, Q., 2010. "Understanding and Thinking for Ideological Emancipation in a New Round of Daqing Oilfield" *Daqing Social Sciences*, 5, 86–87. (in Chinese)

Sun, X.M., Zhou, M., 2011. "The Review and Prospect of Transition to Low-carbon Economy" *Inquiry into Economic Issues*, (6), 116–121.(in Chinese)

Tang, X., Zhang, B. S., Deng, H. M., Feng, L. Y., 2010. "Forecast and Analysis of Oil Production in China Based on System Dynamics" *System Engineering– Theory&Practice*, 30(2), 207–212

Tian, Z., 2008. "The Thinking of World Peak Oil based on High Oil Price" *Technology & Economics in Petrochemicals,* 2(24), 5–10. (in Chinese)

Tian, Z., 2009."Economic Institute.the Market-oriented Process of Energy Industry and its Effects on Energy supply and demand and Carbon Dioxide Emissions" (in Chinese) www.unirule.org.cn/xiazai/2009/20091218.doc

Trends in Oil Prices and Relevant Indices, 2011.*International Petroleum Economic*, 6, 107

Trends of Oil Price and Relevant Indexes, 2010.*International Petroleum Economic*, 2, 85

Trends on oil Prices, 2003. *International Petroleum Economic*, 11, 57

Trends on oil Prices, 2004. *International Petroleum Economic*, 12, 66

Trends on oil Prices, 2005. *International Petroleum Economic*, 13, 65

Trends on oil Prices, 2006. *International Petroleum Economic*, 14, 69

Turner, G. 2008. "A Comparison of the Limit to Growth with Thirty Years of Reality". Available on line: http://www.csiro.au/files/files/plje.pdf

United Nations Development Programme (UNDP), 2010. "China Human Development Report" Published by China Translation and Publishing Corporation, April 2010, ISBN 978-7-5001-2498-6. http://planipolis.iiep.unesco.org/upload/China/China_HDR_2009_2010.pdf

Wan, J.Y. 1994. "Controlled Prediction and an Evaluation System of Oil and Gas 'Resource-Reserve-Production'" *Acta Petrolei Sinica*, 15(3), 51–60. (in Chinese)

Wang, H.M., Xu, X.M., Tian, F., 1960. "Synthetic Oil and Crude Oil Processing Crude Good Start" *Petroleum Processing and Petrochemicals*, 1, 7–8. (in Chinese)

Wang, Y.Z., 1989. "The Several Issues of Gas Development and Utilization in Ancient Sichuang (Social Science Edition)" *Journal of University of Petroleum*, 3, P36–40. (in Chinese)

Wang, D.Z., 2007. "21st Century Outlook of Energy Science and Technology in China" *Beijing: Tsinghua University Press,* 1. (in Chinese)

Wang, N., 2007. "China Oil will be Peak in 2015" *21 shi ji jing ji bao dao*, 10/31, 006. (in Chinese)

Wang, Y., Li, J. W., 2008. "China's Present Situation of Coal Consumption and Future Coal Demand Forecast" *China Pop., Resources and Environ.*, 18(3), 152–155

Wang, T., 2009a."Yumen Oilfield—the cradle of China's Oil Industry" *China Petroleum Daily,* 08. (in Chinese)

Wang, X.Z., 2009b. "Is It Science Or Sensationalization Of Oil Price? Discussion of Peak Oil" *China Economic Weekly*, 34, 26–28. (in Chinese)

Wang, G.Q., 2010a. "The Discovery and Application of oil in Ancient China" *Lantai World,* (19), 15(in Chinese)

Wang, X.Z., 2010b. "Top Ten Oil Fields in China" *China Economic Weekly*, 40. (in Chinese)

Wang, J.L., Feng, L.Y., 2011. "The World's Oil Export Capacity Forecast and Its Impact on China's Oil Import" *Forecasting*, 30(3), 6–10 (in Chinese)

Wang, J.L., Feng, L.Y., Zhao, L., Snowden, S., Wang, X. A comparison of two typical multicyclic models used to forecast the world's conventional oil production. Energy Policy, 2011, 39 (12):7616–7621

Weng, W.B., 1984. "The Foundation of the Forecasting Theory". Beijing: Press of the Petroleum Industry. (in Chinese)

Wu, D.H., 2010. "Peak Oil will be Collapsed without being Attacked" *China Securities Journal*, 6/7, A04. (in Chinese)

Xu D.M., 2009a. "Technology can Decide Future Energy" *China Venture Capital*, 1, 45–48. (in Chinese)

Xu D.M., 2009b. "Insist on Technology Creation and Expend Market" *China Venture Capital*, 4, 6. (in Chinese)

Yanchang oil fields overview. Shaanxi Yanchang Petroleum Co., Ltd (in Chinese). http://www.sxycpc.com/about.jsp?urltype=tree.TreeTempUrl&wbtreeid=1003

Yang, S.B., Zhang, X.L., 1997."The Exploration and Development of Water-Soluble Gas". University of Petroleum press, Dongying. (in Chinese)

Yang, X.H., Gao, W.J., Liu, Y., 2001. "The Study and Perfection of Generalized Models of Oilfield Production Prediction" *Tuha Oil & Gas,* 6(2), 51–55 (in Chinese)

Yang, W., 2010. "Ten Major Oilfields In China 2009" *China Energy News,* 3. (in Chinese)

Yang, M. Y., 2010. "Strategic Trends in Chinese Coal-To-Liquids Development" *Coal Economic Research,* 30(4), 20–24 (in Chinese)

Yu Q.T., 2002. "To Predict Oil Production and Recoverable Reserves of China and USA" *Xinjiang Petroleum Geology,* 23(3), 224–227. (in Chinese)

Yu, T. G., Deng, W. P., 2008. "Forecasting the Increase of Coal Consumption in China Based on the ARIMA Model". *Statistics and Decision,* 24, 89–91 (in Chinese)

Yuan, J. H., Kang, J. G., Zhao, C. H., Hu, Z. G., 2008. "Energy Consumption and Economic Growth: Evidence from China at both Aggregated and Disaggregated Levels" *Energy Economics,* 30(6), 3077–3094

Yuan, N., Ye, K., Mao, J., 2011. "International Oil Prices on the Influence of Biomass Energy in China" *Ecological Economy,* 6, 63–67. (in Chinese)

Zarnikau,J., Guernouche, S., Schmidt, P., 1996. "Can Different Energy Resources be Added or Compared?" *Energy,* 21(6), 483–491

Zhao, X.D., 1987. "The Forecast on Limited Life based on Weng's Model" *Chinese Science Bulletin,* 32(18), 1406–1409. (in Chinese)

Zhang, D.Z., 2000. "50 Years of Technical Progress in Land Petroleum Seismic Exploration" *Oil Geophysical Prospecting,* (6), 683– 694, 702. (in Chinese)

Zhang, S. Y., 2001. "China's Oil Industry in Half of 20th Century" Petroleum Industry press. (in Chinese)

Zhang, B.T., 2002a. "Shan Hai Jing and It's Research Prospects" *Journal Of Yiyang Teachers College,* 23(4), 73–75. (in Chinese)

Zhang, K. 2002b. "Discussion on the Recoverable Gas Resources in China" *Natural gas industry,* 22(6), 6–9 (in Chinese)

Zhang, K. 2003. "Discussion on Recoverable Oil Resources in China" *Oil & Gas Geology,* 24(1), 7–11.(in Chinese)

Zhang, W.Z., 2004. "A Hundred of Practice and a Major Bread through in Chinese Oil Exploration" *China Mining Magazine,* 04, 102–103. (in Chinese)

Zhang, K., 2008a. "Two Prediction of Peak Oil of Hubbert" *Oil Forum,* 6, 22–24. (in Chinese)

Zhang, K., 2008b. "The Analysis of Oil Depletion by Methods of Peak Oil" *Sino-Global Energy,* 13(5), 8–12 (in Chinese)

Zhang, Z. H., Wang, P., 2008. The Impact of The Structure of Primary Energy on Energy Efficiency, Statistics and Decision, 22, 82–83

Zhang, K., 2009. "From Oil Depletion to Peak Oil Theory" *Acta Petroleum Sincia,* 30(1), 154–158. (in Chinese)

Zhao, J. W., Fan, J. T., 2007. "Empirical Research on the Inherent Relationship between Economy Growth and Energy Consumption in China" *Economic Research Journal,* 8, 31–42. (in Chinese)

Zhao, X., 2009. "Will peak Oil Come?" *China Business Times,* 9/7, 002. (in Chinese)

Zhao, Y.S., 2009. "The Reason Analysis of World Production Plateau" *Commercial Times,* 8, 86–92. (in Chinese)

Zhao, Y.S., Zhao, Z., 2009. "Peak Oil: We should Take it Serious" *Research of Economic Theory,* 1, 6–24. (in Chinese)

Zhao, L., Feng L.Y., Hall, C.A.S., 2009. "Is Peakoilism Coming?" *Energy Policy,* 37, 2136–2138

Zhou, Z.Y., Tang, Y.G., 2004. "Current Situation and Problems in China's Oil and Gas Resources Assessment" *Xinjing Petroleum Geology,* 25(5), 554–556. (in Chinese)

Zhou, W., 2009."Development Practice and Regular Exploration of China's Eastern Old Oilfields" *Northeast Asia Petroleum Forum* (in Chinese). Available at: http://www.petroecon.com.cn/2009-dby/pdf/d-2.pdf

Appendix

Table A1 Raw data of direct energy inputs to the oil and natural gas sectors in China

Year	Natural gas	Crude oil	Electricity	Diesel oil	Raw coal	Fuel oil	Gasoline	Refinery gas
Units	10^9 cu.m	10^3 ton	10^9 kWh	10^3 ton	10^3 ton	10^3 ton	10^3 ton	10^3 ton
1995	4.16	1,750	25.9	1,471	2,197	1,664	590	251
1996	3.00	1,691	26.0	1,486	2,267	785	293	65
1997	4.20	3,165	31.5	1,566	3,220	1,109	358	352
1998	3.88	3,125	29.9	1,047	2,164	1,304	329	438
1999	4.56	3,305	30.8	1,453	1,794	1,488	418	562
2000	5.03	4,085	32.2	1,616	1,916	1,505	454	618
2001	5.84	4,227	35.6	1,771	1,701	1,542	436	605
2002	5.93	4,482	36.5	1,976	1,627	1,457	444	635
2003	6.18	5,518	35.7	1,683	1,718	1,249	391	600
2004	4.85	4,990	36.3	1,846	1,703	338	366	372
2005	4.88	5,037	38.5	1,859	1,701	272	257	391
2006	5.46	5,654	31.6	1,874	1,771	289	295	430
2007	6.40	5,695	31.1	1,978	1,745	271	310	333
2008	8.65	6,963	31.8	2,723	1,429	386	278	366
2009	8.91	4,869	33.3	2,302	1,415	270	250	336
2010	10.24	4,823	34.8	1,858	1,627	324	242	356

L. Feng et al., *The Chinese Oil Industry: History and Future*,
SpringerBriefs in Energy, DOI 10.1007/978-1-4419-9410-3,
© Lianyong Feng, Yan Hu, Charles A.S. Hall, Jianliang Wang 2013

Index

L. Feng et al., *The Chinese Oil Industry: History and Future*,
SpringerBriefs in Energy, DOI 10.1007/978-1-4419-9410-3,
© Lianyong Feng, Yan Hu, Charles A.S. Hall, Jianliang Wang 2013